APPLIED MICROBIOLOGY

Trends in Scientific Research 2

Advisory board:

PAOLO BISOGNO, *Italian National Research Council*
SIR JOHN KENDREW, *International Council of Scientific Unions*
ABDUL-RAZZAK KADDOURA, *Unesco*
ALEXANDER KING, *Club of Rome*
IGNACY MALECKI, *Polish Academy of Sciences*
ABDUS SALAM, *International Centre for Theoretical Physics*

Editorial board:

MIKE BAKER, *International Council of Scientific Unions*
AUGUSTO FORTI, *Unesco*
MICHIEL HAZEWINKEL, *Centre for Mathematics and Computer Science*

APPLIED
MICROBIOLOGY

Edited by

H. W. DOELLE

Department of Microbiology, University of Queensland, Brisbane, Australia

and

C.-G. HEDÉN

Microbiological Resources Centre (MIRCEN), Karolinska Institute, Stockholm, Sweden

D. REIDEL PUBLISHING COMPANY

A MEMBER OF THE KLUWER ACADEMIC PUBLISHERS GROUP

DORDRECHT / BOSTON / LANCASTER / TOKYO

UNESCO, PARIS

Library of Congress Cataloging in Publication Data
Main entry under title:

Applied microbiology.

  (Trends in scientific research ; 2)
  Includes bibliographies and index.
  1. Industrial   microbiology. 2. Biotechnology.
I. Doelle,   H.   W.   II. Hedén,   Carl-Göran. III.
Series.
QR53.A66   1985   660'.62   85–31185
ISBN 90–277–2095–9
ISBN 92–310–2335–7 (Unesco : pbk.)

Published by the United Nations Educational,
Scientific and Cultural Organization,
7 Place de Fontenoy, 75700 Paris, France
and D. Reidel Publishing Company,
P.O. Box 17, 3300 AA Dordrecht, Holland

Hardbound edition sold and distributed in the U.S.A. and Canada
by Kluwer Academic Publishers, 190 Old Derby
Street, Hingham, MA 02043, U.S.A.

Hardbound edition in all other countries, sold and distributed
by Kluwer Academic Publishers Group,
P.O. Box 322, 3300 AH Dordrecht, Holland

D. Reidel Publishing Company is a member of the Kluwer Academic Publishers Group

Limpbound edition sold throughout the world by
Unesco, 7 Place de Fontenoy, 75700 Paris, France

The designations employed and the presentation of the material in this publication
do not imply the expression of any opinion whatsoever on the part of the publishers
concerning the legal status of any country or territory, or of its authorities, or
concerning the frontiers of any country or territory.

The authors are responsible for the choice and the presentation of the facts contained
in this book and for the opinions expressed therein, which are not necessarily those
of Unesco and do not commit the organization.

Printed in The Netherlands

Contents

Preface

Unesco, in cooperation with the International Council of Scientific Unions (ICSU), and a number of international and national institutions, such as the International Federation of Institutes for Advanced Study, is publishing a series of monographs 'Trends in Scientific Research'. Each monograph will be complete in itself, but will be linked to the others by a central theme — that of developments at the forefront of science of special interest to humankind.

The idea is far from new. Professor P. Auger published a major work in 1957, *Trends in Scientific Research*, which played a seminal role in stimulating the development of science in many countries and served for many years as a handbook for science policy makers. Since this book was published, there has been a considerable extension of our scientific knowledge and it is no longer possible to fit all of science into one book if the various disciplines are to be treated in some depth. It was therefore decided to publish a number of separate monographs written by specialists and linked together to provide an extensive and extending image of present and future trends in science.

The monographs will be interdisciplinary in character, bring together not only research in several disciplines but also the applied as well as the basic aspects of science, and in particular the uses of science in forecasting and in problem solving. Special attention will be paid to the use of science in the fulfilment of human needs.

The monographs are addressed to decision makers, to scientists interested in disciplines other than their own, the educated man in the street, and so on. It is intended that the series will provide interesting material that will influence young students to take up a career in the subjects treated.

Inasmuch as there will be a focus on applied aspects of the subjects treated, the monographs will also be of use in highlighting those types of scientific research most relevant to the requirements of developing countries and those fields of research and developments with the

greatest potential applications. Moreover, they will also try to highlight and foresee the most important aspects of the consequences on society of the introduction of new technologies.

The monographs will provide a series of interesting and stimulating publications which, it is hoped, will play an important role in shaping the development of scientific research and stimulating a greater interest in science in the world at large.

THE EDITORIAL BOARD

List of Contributors

Chiao, J. S., Department of Microbiology, Shanghai Institute of Plant Physiology, Academia Sinica, Shanghai, China.

DaSilva, E. J., Division of Scientific Research and Higher Education, Unesco, Place de Fontenoy, 75700 Paris, France.

Doelle, H. W., Biotechnology Group, Department of Microbiology, University of Queensland, St. Lucia, Queensland 4067, Australia.

Gyllenberg, H. G., Department of Microbiology, University of Helsinki, Helsinki 71, Finland.

Hedén, C.-G., UNEP/Unesco/ICRO Microbiological Resources Centre (MIRCEN), Karolinska Institute, 19401 Stockholm, Sweden.

Heinritz, B., Institut für Technische Chemie, Akademie der Wissenschaften der DDR, Leipzig, German Democratic Republic.

Olguín, E. J., Instituto Mexicano de Tecnologías Apropriadas, Apartado Postal 63–254, 02000 México D.F.

Ringpfeil, M., Institut für Technische Chemie, Akademie der Wissenschaften der DDR, Leipzig, German Democratic Republic.

Sly, L. I., Culture Collection, Department of Microbiology, University of Queensland, Brisbane, Australia.

Zavarzin, G. A., Institute of Microbiology, USSR Academy of Science, Puschino, USSR.

Introduction

Biotechnology has been defined as *'the application of scientific and engineering principles to the processing of materials by biological agents to provide goods and services'* (Bull *et al.*, 1982). This definition implies that applied microbiology is an integral part of biotechnology. The term 'scientific and engineering principles' further indicates that subjects such as microbial technology, biochemical engineering, microbiology, biochemistry and genetics form the core of biotechnology. 'Biological agents' can include microorganisms, plant cells, animal cells or enzymes, and similarly, 'materials' all organic and inorganic substances. There is no doubt that biotechnology is an area which cuts right across the traditional faculties of biology, engineering and medicine and thus is characterized by its interdisciplinary nature. It is true that industrial microbiology and food fermentation are as old as mankind, but both represent only a portion of what is now referred to as biotechnology, although both terms have been employed with different connotations by politicians and project planners (DaSilva, 1983). The introduction of economics and socio-economics as integral parts of biotechnology makes this field suitable for both rural and industrial applications (Rivière, 1983).

The greatest challenge in the field of biotechnology has developed without any doubt over the past few decades. The realization of possible food shortages in the near future, disasters in climatic conditions leading to starvation in many areas of the world, the accumulation of industrial, human and animal wastes as a consequence of an ever-increasing population, together with an awareness of the changes in the environment that are increasing health hazards, have alarmed both scientific and governmental agencies. The energy crisis brought the realization that fossil fuels, up to now, have been channelled primarily towards meeting the population growth rate and its technological accomplishments (DaSilva and Doelle, 1980). This sudden realization of how much our lives depend upon fossil fuel energy and

the fact that our natural fossil fuel resources will diminish with the ever-increasing demand by the growing population instigated a desperate search for resources that are renewable and technologies that could be used to replace the heavy energy-demanding industries. It became clear to the scientific and governmental communities that goods and services at present provided by the chemical industries must be partly or wholly replaced by those from biological industries, if mankind is to survive.

This realization immediately divided biotechnological development into two large camps, which coincide with development in the developed and the developing countries. Whereas the former is mainly concerned with maintaining the quality of life, the developing countries' main thrust is towards the improvement of the quality of life (Natl. Acad. Sciences, 1979; van Hennert *et al.*, 1983). Starved of industrial development in the past, the developing world sees the chance of avoiding the establishment of energy-demanding industries and of promoting the use of renewable resources via biological industries (Campos-Lopez, 1980; King *et al.*, 1980), of facilitating the evolution of relevant technologies that can serve as the base for industrial growth and employment, and also of seeing economic growth in urban and rural areas through the just distribution of food, fuel, and the like (DaSilva 1983). There is an important consideration here: that a certain development may be appropriate in one situation, but unsuitable to a variable degree in another. The reasons for this are manifold, including the availability of different renewable resources, a different socio-economic order, lack of infrastructure and like. A new term was created for this development: *'appropriate biotechnology'*. In order to develop such appropriate biotechnology, careful consideration must be given to the resources available together with the existing social structure of the population (Doelle, 1982). The motivation for any developmental effort in this area should be either to satisfy basic needs, such as food, shelter and living conditions or to improve standards (Sørensen, 1979). The idea that improvement of the situation in the developing countries can be achieved through simple transfer of this 'appropriate biotechnology' or 'developed countries' technology' is therefore a naive and fruitless one. One needs to study the general technology available in other countries and to choose or select that process which could be best modified and with the available manpower to suit the local requirement. It is encouraging to see a definite trend towards such deve-

lopment by scientists of the developing countries. The best examples have been the attempts to integrate biotechnology with agricultural systems (Monroy and Viniegra, 1981; Olguin, 1982) in order to completely exploit and secure the renewable resource material, resulting not only in a range of different goods but at the same time in a cleaner and better environment.

The challenge lies in the difficulties which exist in biotechnological development. Taking into account that considerations are the existence of basic processes that can be geared towards the development as well as infrastructures of the particular country, be it developed or developing. Consequently, an awareness of the trends in the field of biotechnology is a great asset.

The two different directions of biotechnological development are reflected at the international level by the existence of the International Biotechnology Symposium devoted primarily or predominantly to the developed countries and the Global Impact of Applied Microbiology (GIAM) devoted to the developing countries. It is the function of the International Organization of Biochemical Engineering and Biotechnology (IOBB) in cooperation with the International Cell Research Organization (ICRO) to form a bridge between the two camps, and it was heartening to witness the fruitful interconnections as they occurred at the VIIth International Biotechnology Symposium in New Delhi in 1984.

There is no doubt that the social and economic pressures stemming from an increasing demand for better health and nutritional standards, environmental concerns for waste management, and the increasing cost of oil-based feedstocks, will ensure that biotechnology has a major impact in the world (Bull *et al.*, 1982; Healy, 1979). Some of these pressures are important in the development of appropriate socioeconomic bioindustries (DaSilva, 1980). DaSilva (1981) used the terms high-capital, intermediate-capital and low-capital or village-type technology. Some village-type technology is, of course, as old as industrial microbiology. Here we are referring to the solid substrate or food fermentation industry (Djien, 1982). Whereas industrial microbiology has made spectacular advances, the food fermentation industry has not developed as quickly. The reasons are obvious if one concerns oneself with the process technologies. It is much easier to develop processes with a single pure culture than with a mixed culture population.

This brings us to the heart of the problems and the processes and major concern in the development of biotechnology: the catalyst or biological agent itself. If we restrict ourselves to the microorganism as the biological agent, the enormous potential of this biological agent has been stressed many times in the past (Porter, 1980; DaSilva 1981; Bull *et al.*, 1982). The problem is, however, that the number of microbiologists who have the understanding and vision to convince the public that the beneficial activities of the microbial world can be exploited for the human good, is still very small. Process-oriented basic research is still relatively weak, as it requires an acknowledgement by the public and research bodies that such a direction is at least as useful as basic research for the advancement of pure science (Bull *et al.*, 1979). One has only to mention the developments in plant cell and animal cell cultivation, hybridoma technology, genetical engineering, immobilization and the like, to visualize the benefits coming from biotechnology.

This volume on trends in applied microbiology and biotechnology has as its aim to show the achievements, trends and future scope of biotechnology in developing countries. The enormous range of R&D work undertaken in the world at present forces the editors to be selective in their approach and it is hoped that a future volume in the series might cover other areas as they develop further and show definite trends.

Microbial cultures or their enzyme components are the basic elements in applied microbiology and biotechnological processes, as they represent the catalysts for these processes. Most of us have used cultures from culture collections, but are unaware of their ever-increasing importance in preserving the microbial culture heritage. The first chapter discusses the World Data Centre and Microbial Resource Centres (MIRCENs) established by Unesco and UNEP for the purpose of providing the manpower skilled in the preservation of new cultures and acting as the data source for the location and supply of particular cultures. It should be realized that maintenance and preservation depend on the type of organism and become almost an art if they concern the viability of the microbe and its capability in a microbial process. Many cultures have been lost in the past owing to mismanagement and we cannot afford the enormous efforts required for the development of new strains (Rivière, 1975), and particularly patented strains. There is no doubt that development in biotechnology

depends very much on future trends in microbial process development, which involves isolation, optimization and preservation of our cultures. It is very unfortunate that many government agencies often overlook the importance of culture collections and it is hoped that the MIRCEN network together with the new *MIRCEN Journal of Applied Microbiology and Biotechnology* will rectify the situation in the near future and safeguard new biotechnological development.

Biotechnology is a field that relies very heavily on the information content of the microbe (Bull *et al.*, 1982), which has a vast potential to be tapped (DaSilva and Doelle, 1980). The exploration and search for this information and its exploitation for the development of a microbial process is outlined in the second chapter. Very often overlooked by many researchers is the potential of the right microbe. If this is the case for the so-called wild-type strains, where are the limits if one considers the opportunities of improvement and extension created through genetical engineering in conjunction with the new cultivation techniques developing through the immobilization technology?

The greatest benefit to the whole of microbial technology comes from the development of the single-cell protein (SCP) process (Bull *et al.*, 1979; Rose, 1979). Technological innovations in the handling of water-insoluble substrates, pressure-cycle fermenter design, the first large-scale chemostats, and tower fermenter desgin are a few examples of the initial concepts coming from the biochemical engineers that have stimulated interest and activity in this field. It is often forgotten that the initial work with n-alkanes as substrate was carried out because n-alkanes were unwanted or waste products of oil refineries, as were also in many cases methanol and methane. In order to bring unit costs down, the oil industry had to find new products from such wastes. The SCP process was certainly the first attempt to produce proteinaceous food and feedstuffs without the aid of agriculture and independent of climatic conditions. M. Ringpfeil and B. Heinritz describe trends in single cell protein technology as SCP is an ever-increasing source of animal feed and can now be produced using a large variety of waste products.

Apart from food and feed, many areas in the world are still without energy and electricity for cooking and industrial development, and others feel the high prices of oil imports. The oldest and most advanced technology can be found in the conversion of domestic and

agricultural wastes into methane (Hobson *et al.*, 1981). Although the basic foundations for biogas reactor development were laid in India, the most spectacular advance in their exploitation have occurred in the People's Republic of China. This enormous development is outlined in the chapter by J.S. Chiao.

E. J. Olguin then gives an example of the future trends in socio-economic or appropriate biotechnology. This integrated system combines the potential of algae, yeast and bacteria with agricultural systems and waste treatment technology to raise not only the economic, but also the health and nutritional standard of the people in a small community using typical village-type technology. Such integrated systems can lead to food, feed and fertilizer supply, can take care of environmental problems and certainly raise living standards in arid zones. Many of these systems have been proposed not only in Latin America, but also in Kuwait and southeast Asia.

The following two chapters by Zavarzin and Heden - are significant in that they outline two important aspects that are apt to be overlooked in biotechnological approaches to development. Whereas the former outlines some of the biochemical and ecological aspects that may impinge on new bio-industries such as ore-leaching, the latter focusses on the paradoxical scale-factor in biotechnology. In this chapter attention is also given to the important aspect of biotechnology transfer.

The final chapter deals with a number of biotechnological considerations that interrelate with world development. In a discussion on policies, potentials and prospects a number of cases are examined in the light of biotechnology being deployed for goods and services. In agreeing that "known" goods and services can be produced more beneficially than before, the conclusion dervied at, is that real and far-reaching progress is bound actually to *new* products, and completely *new* goods and services. These, hardly foreseen, are not outside the scope and reach of biotechnology.

It is the aim of this volume to indicate trends in applied microbiology using the most established systems as examples whether complete or still in the process of development. There is no doubt that much exciting developmental work is in progress in many other areas, but it is a long way from an idea, via development, to established reality. If this volume succeeds in passing on some of the excitement about the work that has already been done using the vast potential of

the biological agent, indicates what could be done if this potential is harnessed further, and stimulates even stronger development, then its purpose will have been achieved to a great extent.

REFERENCES

Appleton, J. M., V. F. McGowan and V. B. D. Skerman (1980). *Microorganisms and Man.* Unesco/UNEP Contract No. 258117.

Behbehani, K., I. Y. Hamdan, A. Shams and N. Hussain (1980). 'Bioconversion Systems for Feed Production in Kuwait.' In *Bioresources for Development* (A. King *et al.*, eds.), pp. 159–171. Pergamon Press.

Bull, A. T., D. C. Ellwood and C. Ratledge (1979). *Microbial Technology: Current State, Future Prospects.* Soc. Gen. Microbial. Symp. **29**. Cambridge University Press.

Bull, A. T., G. Holt and M. D. Lilly (1981). *Biotechnology.* OECD Paris.

Campos-Lopez, E. (ed.) (1980). *Renewable Resources: A Systematic Approach.* Academic Press Inc.

DaSilva, E. J. (1981). 'The Renaissance of Biotechnology: Man, Microbe, Biomass and Industry.' *Acta Biotechnologica* **1**, 207–246.

DaSilva, E. J. (1983). 'Biotechnology in Development of Cooperation: A Developing Countries View.' In *Biotechnology in Developing Countries'* (P. A. van Hemert *et al.*, eds.). Delft University Press.

DaSilva, E. J. and H. W. Doelle (1980). Microbial Technology and Its Potential for Developing Countries. *Process Biochem.* **15**(3), 2–6.

DaSilva, E. J. (1980). 'Trends in Microbial Technology for Developing Countries.' In *Renewable Resources: A Systematic Approach.* (E. Campos-Lopez, ed.) Academic Press Inc.

Doelle, H. W. (1982). 'Appropriate Biotechnology in Less Developed Countries. *Conservation & Recycling* **5**, 75–77.

Doelle, H. W. (1983). 'Biological Constraints in Solid Substrate Fermentation.' In *Seminar on Solid Substrate Fermentation* held at Cebu City (Phillippines) in August 1983 by ASEAN-Protein Project.

Healy, A. T. A. (1979). *Science and Technology for What Purpose? An Australian Perspective.* Austral. Acad. Science, Canberra.

Hobson, P. N., S. Bousfield and R. Summers (1981). *Methane Production from Agricultural and Domestic Wastes.* Appl. Science Publ. Ltd.

King, A., H. Cleveland and G. Streatfield (eds.). *Bioresources for Development: The Renewable Way of Life.* Pergamon Press, New York.

Monroy, O. and G. Viniegra (1981). *Biotechnologia para el aprovechamiento de los desperdicios organicos.* AGT Editor, S. A. Mexico.

National Academy of Sciences (1979). *Microbial Processes: Promising Technologies for Developing Countries.* National Acad. Sciences, Washington.

Olguín, E. (1982). 'Conversion of Animal Waste into Algae Protein within an Integral Agriculture System. In *Microbiological Conversion of Raw Materials and By-products of Agriculture into Proteins, Alcohols and Other Products.* Seminar held in Novi Sad, Yugoslaria, June 1982.

Rose, A. H. (1979) (ed.). 'Microbial Biomass.' *Economic Microbiology* **4**, Academic Press Inc.
Rose, A. H. (1982) (ed.). 'Fermented Foods.' *Economic Microbiology* **7**, Academic Press, Inc.
Rivière, J. la (1975). *Industrial Applications of Microbiology*. Masson et Cie, Parîs.
Rivière, J. la (1983). 'Biotechnology in Development of Cooperation: A Donor Countries View. In *Biotechnology in Developing Countries*. (P. A. van Hemert *et al.*, eds.). Delft University Press.
Sørensen, B. (1979). 'Energy Technology and Social Structure.' In *Appropriate Technology for Underdeveloped Countries*, pp. 38−55. Second Intern. Symp. Engineering, San Salvadore.

H. W. DOELLE
C.-G. HEDÉN

L. I. Sly

Culture Collection Technologies and the Conservation of Our Microbial Heritage

INTRODUCTION

Microbial cultures or their enzyme components are the basic elements in applied microbiology and biotechnological processes. The diversity of the microbiological community offers a seemingly limitless pool of metabolic information for the production of useful products or the conversion of others for the benefit of mankind. Technological development in the science of genetic engineering has brought a new perspective to our microbial heritage and brought many processes within the practical reach of the biotechnologist.

If Antonie van Leeuwenhoek was the first microbiologist, then Louis Pasteur should surely be regarded as the first applied microbiologist. It was he who played the important fundamental role in recognizing the link between microorganisms and the processes now regarded as applied microbiology and biotechnology. Pasteur observed that microorganisms had a specific role in human and animal diseases, in the production of foods and drinks such as cheese, beer, wine and vinegar, and in production difficulties such as those in the silk industry.

The early observations by Pasteur and other microbiologists confirmed the close association between microorganisms and mankind and showed that these associations could be either beneficial or harmful to man. The subsequent isolation of pure cultures by Koch in 1880 was soon followed by others and the concept of the collection of cultures was born. Over the last 100 years microbiologists have been active in isolating pure culture strains of microorganisms which comprise the species and genera of the microbial world. This task is not complete by any means and perusal of the scientific literature shows that many new taxa are described each year.

The isolation of pure cultures is an essential part of the identification of microorganisms in nature and the selection of microorganisms which exhibit rare or unusual capabilities of degradation or production of compounds which could be exploited by man.

1

It is well accepted that microorganisms play a vital and indispensable role in the recycling of nutrients in nature. Man's association with selected environmental microorganisms has developed over the years so that various facets of applied microbiology regularly affect our daily lives and our dependence on these processes is rapidly increasing and will no doubt continue to do so in the future.

Microbial cultures are now used in medical and veterinary pathology; in industrial fermentation in the chemical, brewing and pharmaceutical industries; in agriculture, biological nitrogen fixation, plant pathology and dairying; in food technology and biotechnology; in vaccine production; and in environmental and ecological studies.

Also, many industries are using microorganisms as biological "tools" in the production of goods or in testing product quality, and in food preservation; in addition, many institutions require internationally known cultures for strain diagnosis of pathogens in human, plant and animal health and in numerous taxonomic and ecological studies.

All these areas of microbiology have had and will continue to have a basic requirement for the supply of authentic cultures of microorganisms, and the increasing importance of culture collections has been accepted as a growing number of microbiologists and biotechnologists are faced with situations requiring microbial cultures.

Pure and selected cultures of microorganisms have taken food manufacture beyond an art into one of a controlled process. This change, although well established in the industrially developed nations, is only just beginning in many developing nations. Culture collections with properly preserved starter cultures form the basis for the change which will help to ensure the maintenance of a stable food supply and the lessening of waste of valuable food resources due to failed fermentations.

The rising cost of chemical fertilizers has focussed increased attention on the role of microorganisms in biological nitrogen fixation and the benefits of legume inculation with host-specific strains of *Rhizobium* are being exploited with renewed vigour. The challenge of new environments, new hosts and new *Rhizobium* strains is being met in developing tropical regions and once again emphasizes the need to ensure that this world resource selected from the environment is properly conserved.

Many biotechnological industries such as alcohol fermentations and antibiotic production are based on the exploitation of a single microbial strain. The high cost of strain selection, research and development,

patent applications and product promotion and acceptability means that there is a considerable capital investment involved. This, coupled with the fact that the economic viability of some biotechnological industries may be only marginally superior to alternative processes, underlines the need to maintain the productive efficiency of the strains which form their basis. On the other hand, the protein-rich food supply of a family or village enterprise in a developing country may depend on the efficient growth of a starter culture for an indigenous fermented food selected over many years.

Once isolated, it is essential that microbial strains are conserved in a genetically stable condition. There is amongst microorganisms an inherent tendency to mutate, which may be accelerated in the artificial laboratory environment. The need for adequate preservation methods became obvious in the very early days of microbiology. It became evident that changes in colonial morphology were associated with changes in antigenic patterns and that loss of virulence also occurred after prolonged laboratory subculture.

Such changes in antigenic structure or metabolic activity may be helpful to the applied microbiologist seeking to develop new processes or enhance yields, but once a strain is selected for a process it is essential that the strain is properly preserved in a stable and reliable way. Culture collections fulfill this important role by ensuring that representatives of as many microbial types which have been isolated are adequately conserved for the present and future generations.

The selection of microbial cultures and their preservation in culture collections is a continuing process. A recent survey of 356 culture collections in 52 countries (Tables I and II) by the World Data Centre for Microorganisms (McGowan and Skerman, 1982) revealed that there are well over half a million cultures preserved in the world's culture collections, an increase of some 40% in the last decade. It is fair to say that some of these cultures are duplicated in a number of collections because they are universally used for comparative purposes or in internationally accepted methods of analysis, but by and large the majority of these cultures represent an enormous gene pool of regional and international interest.

This massive collection of strains includes representatives of most of the types of microorganisms that have been successfully isolated to date. This world wide collection contains in the order of 12,500 species or varieties of filamentous fungi, 3500 bacteria, 2000 yeasts, 2000 algae,

TABLE I

Culture collections in various regions of the world (McGowan and Skerman, 1982)

Region	No. of countries	No. of collections	University	Sponsor Private industry	Government
Africa	5 (9)	10[a] (11)	3 (4)	1 (2)	7 (5)
Asia	9 (10)	62[b] (30)	28 (10)	8 (2)	32 (18)
Australia — New Zealand — Papua New Guinea	3 (2)	51[c] (39)	11 (9)	5 (5)	37 (25)
Europe (Eastern) & USSR	7 (7)	39[d] (38)	12 (10)	5 (0)	27 (28)
Europe (Western)	16 (12)	107[e] (100)	45 (39)	17 (28)	49 (33)
Latin America	7 (6)	23[f] (23)	12 (12)	2 (0)	13 (11)
Middle East	3 (6)	5[g] (13)	2 (8)	0 (1)	3 (4)
North America	2 (2)	59[h] (67)	18 (28)	12 (15)	34 (24)

( ) Data from Martin and Skerman (1972) (Porter, 1976).

[a] Nigeria 2, Republique du Senegal 1, South Africa 3, Uganda 2, Zimbabwe 2.

[b] China 1, India 25, Indonesia 2, Japan 11, Malaysia 4, Philippines 7, Singapore 1, Sri Lanka 2, Thailand 9. (Sponsorship not declared by 3 collections).

[c] Australia 41, New Zealand 9, Papua New Guinea 1. (Sponsorship not declared by 1 collection).

[d] Bulgaria 2, Czechoslovakia 13, Hungary 9, Poland 6, Rumania 2, Union of Soviet Socialist Republics 6. (Sponsorship not declared by 3 collections).

[e] Austria 2, Belgium 3, Denmark 6, Federal Republic of Germany 14, Finland 2, France 15, Greece 2, Ireland 2, Italy 11, Norway 3, Portugal 1, Spain 2, Sweden 1, Switzerland 3, The Netherlands 4, United Kingdom 36. (Sponsorship not declared by 4 collections).

[f] Argentina 6, Brazil 8, Chile 2, Colombia 1, Ecuador 1, Mexico 2, Venezuela 3. (Sponsorship not declared by 1 collection).

[g] Israel 2, Jordan 1, Turkey 2.

[h] Canada 30, United States of America 29. (Sponsorship not declared by 1 collection)

300 protozoa, 38 lichens, 1850 animal viruses, 180 bacterial viruses, 43 insect viruses and 306 plant viruses.

Apart from the obvious scientific and industrial importance of these cultures they represent an enormous financial investment by goverments and industry. Based conservatively on 100 hours salary required for the isolation, characterization and preservation of the half million strains, this world asset is valued at 500 million dollars and with other overheads and research and development costs this value could well

TABLE II

Number of collections in the world listing various
holdings (McGowan and Skerman, 1982)

Kind of holding	No. of collections	
Algae	49	(18)
Bacteria	261	(170)
Filamentous fungi	162 ⸴	(126)
Lichens	3	(2)
Protozoa	28	(16)
Tissue cultures	31	(4)
Viruses		
Animal	54	(30)
Bacterial	45	(22)
Insect	6	(3)
Plant	8	(6)
Yeasts	125	(100)

(   ) Data from Martin and Skerman (1972) (Porter, 1976).

by multiplied by a factor of 10 or even 100. An investment of this magnitude is deserving of the most conscientious support.

Just under 1000 people are engaged in full-time employment in culture collections (McGowan and Skerman, 1982) including 774 full-time professional and technical staff and an additional 70 equivalent full-time positions contributed through part-time assistance. On average each laboratory staff member is responsible for some 670 cultures. This figure by no means reflects the actual situation in all collections, as many are considerably better or worse off than this position.

The data in Table I show that culture collections are well distributed throughout most regions of the world and that there has been a growth of some 11% in the collections registered with the World Data Centre for Microorganisms (WDC) over the past decade. However, although there has been an overall growth of 11% in the number of collections registered, 119 collections registered in 1972 have not been registered in 1982, and just what arrangements have been made to conserve any valuable cultures in these collections is not known. One encouraging

sign though is the increase from 30 to 62 in the number of collections in Asia. This increase reflects the increased awareness of culture collections in Asia, principally in relation to the cultures involved with indigenous fermented foods and biological nitrogen fixation. One cause for concern is the apparent decline in the number of collections in the Middle East.

The data in Table I also show that the majority of culture collections are sponsored by universities or governments. Only 50 collections receive private or industrial funding and of these only 27 or less than 8% of the world's culture collections are funded from private sources alone. This reflects the position to date that only a few restricted areas of microbiology which are profit-earning require the services of a culture collection of a substantial nature. Industrially funded culture collections are often associated with companies involved in the selection of large numbers of strains for the production of novel substances such as antibiotics. Of the eight privately funded collections in Asia five are located in Japan. This is a reflection of the profitability of fermentation industries in Japan and the contribution they make to that country's income. Approximately 37% of culture collections are associated with universities, which gives a clear indication of the value of culture collections in education and research. Governments directly fund in full or part 57% of culture collections. This commitment by government indicates the value of collections in government service in human and animal health, agriculture and related areas, and research and education.

The main interests of culture collections are summarized in Table III. It is interesting to note that, contrary to the common impression that medical microbiology is the dominant interest of microbiologists, only 33% of culture collections declared medical microbiology as a major interest, and an equal number of collections declared agriculture and forestry as their main concern. Overall, 61% of registered collections indicated that applied microbiology was their main interest and covered areas such as agricultural and plant pathology, forestry, industrial, biodeterioration, dairy, food and geomicrobiology. The upsurge in interest and work in the field of genetics is reflected in the fact that five collections now specifically list this as one of their main interests.

There has been an increase of approximately 40% in the cultures held in culture collections over the past ten years. Table IV shows that holdings for all types of cultures have increased substantially during this period and that the increases have occurred in all regions of the world except the Middle East.

TABLE III
Main interests of various collections in the world
(McGowan and Skerman, 1982)

Microbiological interest	No.	
Agricultural	100	(80)
Forest products	20	(11)
General	136	(144)
Industrial	83	(83)
Insect	14	(5)
Medical	119	(121)
Veterinary	48	(20)
Other main interests		
Assay	0	(1)
Biodeterioration	3	(1)
Dairy	4	(0)
Educational supplies	2	(0)
Food	3	(0)
Genetics	5	(0)
Geomicrobiology	2	(0)
Marine microbiology (including algology)	7	(0)
Plant pathology	3	(3)
Taxonomic mycology	0	(1)

(   ) Data from Martin and Skerman (1972) (Porter, 1976).

The historical development of culture collections has been reviewed by Porter (1976). The case for culture collections, the important fundamental role they play in all aspects of microbiological endeavour, and their usefulness in developing and developed regions of the world, have all been the subject of papers and discussions at a number of international conferences and specialist meetings (Martin, 1963; Iizuka and Hasegawa, 1970; Pestanade Castro *et al.* 1973; Fernandes and Costa Pereira, 1977 and Rogosa, 1981).

The establishment, development and continuity of culture collections were the subjects of resolutions coming from the United Nations Conference on the Human Environment in Stockholm in 1972. It was this meeting which reached agreement on an international programme to preserve the world's genetic resources and gave rise to the United Nations Environment Program (UNEP) (Martin, 1976).

TABLE IV

Number of holdings in various collections in the world by region (McGowan and Skerman, 1982)

Region	Algae	Bacteria	Filamentous fungi	Lichens	Protozoa	Tissue culture lines	Viruses				Yeasts	Total for Region
							Animal	Bacterial	Insect	Plant		
Africa	0 (19)	3,440 (1,917)	28 (340)		0 (812)	10	608 (226)			48 (1)	9 (51)	4,143 (3,366)
Asia	1,139 (468)	25,997 (14,589)	20,584 (11,748)	100	18 (30)	146 (30)	618 (597)	919 (412)	20	10	6,622 (5,117)	56,173 (32,991)
Australia – New Zealand – Papua New Guinea	149 (39)	20,772 (6,902)	13,796 (7,464)		6 (2)	219 (2)	1,445 (226)	530 (239)	26	6 (52)	820 (325)	37,769 (15,251)
Europe (Eastern) & USSR	1,022 (1,058)	25,670 (22,208)	7,328 (7,115)		10 (25)	17 (10)	981 (817)	290 (221)		27	8,291 (5,000)	43,636 (36,454)
Europe (Western)	7,423 (3,926)	91,054 (76,084)	53,836 (29,284)		2,129 (345)	177	2,677 (667)	930 (304)	81	226	14,790 (15,551)	173,323 (126,161)
Latin America	13 (12)	12,540 (8,168)	2,144 (1,803)	2	4 (1)	7 (10)	48 (110)	8 (26)	2		1,790 (2,573)	16,558 (12,703)
Middle East		1,872 (8,502)	165 (85)		8 (7)	3	52 (40)	160 (54)			199 (107)	2,459 (8,795)
North America	785 (232)	147,236 (106,524)	63,104 (52,327)		790 (87)	442 (120)	2,048 (1,295)	530 (569)	18	204 (173)	19,702 (10,202)	234,859 (171,529)
Total for kind of holding	10,531 (5,754)	328,581 (244,894)	160,985 (110,166)	102	2,965 (1,309)	1,021 (172)	8,477 (3,978)	3,367 (1,825)	147	521 (226)	52,223 (38,926)	568,920 (407,250)

( ) Data from Martin and Skerman (1972) (Porter, 1976).

The sentiments of this resolution were reinforced recently at the National Work Conference on Microbial Collections of Major Importance to Agriculture in 1980, which advised that "the need for good management in preserving representative, type, and industrially or agriculturally useful strains of microorganisms and germ plasm is urgent" (Rogosa, 1981).

Over the years biosynthetic products have received particular attention in applied microbiology and encouraged the growth of industrial microbial collections. The strains in these collections contain an important part of the gene pool of our natural microbial heritage. As the age of genetic engineering dawns, culture collections have never been more deserving of consideration in the world's scientific, industrial and social development as the significance of a broad ranging representative microbial gene pool becomes more obvious and important to us all. Rare properties once thought to be trapped inside industrially unsuitable strains of microbial or other life forms may be exploited to the full advantage of mankind.

The microbial gene pool conserved in culture collections is a world resource, the significance of which may only be recognized in the light of future scientific discoveries. It is essential then that these important cultures are preserved adequately and remain as genetically stable as possible. Culture collections have developed management procedures and preservation techniques which are aimed at fulfilling this responsibility with which they are charged. The remainder of this chapter will elaborate on the general operational guidelines and procedures followed in culture collections. Each collection, however, must develop its own system, according to institutional demands, the nature of the cultures held and the equipment available.

THE FUNCTIONS OF CULTURE COLLECTIONS

Culture collections are permanent laboratories where important microbial cultures are collected and preserved and made available on demand. The prime functions of a culture collection are the deposition, preservation and distribution of cultures of microorganisms. A culture collection therefore acts as a repository for the safe storage of microorganisms and for the preservation of their genetic stocks.

Culture collections may be divided into three main types. They are service collections, institutional collections and private collections.

Service collections are limited in number and are permanent resource collections which usually have national status in terms of funding. These collections usually attempt to assemble a broad collection of cultures without holding a large number of strains of each type. The accessioning policy of each collection is different and may vary from a comprehensive range of all types of microbial cultures such as in the American Type Culture Collection, to say a more limited range of a single type such as the bacteria in the National Collection of Plant Pathogenic Bacteria, or in the National Collection of Industrial Bacteria which excludes pathogenic bacteria.

Service collections are a truly available world as well as national resource in that this type of collection supplies cultures on demand to the scientific community at large. These collections may have been established primarily for this purpose, or may have evolved to this role from being originally an institutional collection whose holdings became well known and sought after. Some service collections still maintain their institutional role as well. Service collections usually charge a fee for cultures and publish comprehensive catalogues. Some examples of service collections are listed in Table V.

Institutional collections are usually associated with departments or institutions which have a charter to provide a microbiological service of some kind. The service role may range from education and research in a university, to support of an identification service in a government department or hospital, to research and development in industry. Service and institutional collections of importance in applied microbiology and biotechnology have been listed by Hesseltine and Haynes (1973) and McGowan and Skerman (1982). The holdings of institutional collections may range from being even more diverse than some service collections to specialized medical, veterinary, agricultural or industrial microorganisms. The holdings may cover a range of microbial types or be limited to a single type such as the bacteria, fungi, algae or viruses. Many institutional collections are internationally known for their expertise in the identification of particular types of microorganisms. This expertise is often recognized by designation as a reference centre by the World Health Organization (WHO) or other bodies.

Private culture collections are usually personal research collections. Often these collections have no guaranteed permanency beyond the working life of the microbiologist actively working with the collection. Such collections may be associated with a lifetime of research or with an

TABLE V

Examples of service collections in various countries and registered with the World Data
Centre (WDC)

Country	WDC collection no.	Service collection[a]
Czechoslovakia	65	Czechoslovak Collection of Microorganisms (CCM)
Federal Republic of Germany	274	Deutsche Sammlung von Mikroorganismen (DSM)[b]
Japan		Japanese Federation of Culture Collections (JFCC) — coordinating organization for 22 collections including:
	190	Institute of Applied Microbiology, University of Tokyo (IAM)
	191	Institute for Fermentation, Osaka (IFO)
	195	Laboratory for Fermentation, Hiroshima University (HUT)
	208	Institute of Medical Science, University of Tokyo (IID)
	219	Department of Fermentation Technology, Osaka University (OUT)
	235	Research Institute of Fermentation, Yamanashi University (RIFY)
	243	National Institute of Agricultural Sciences (NIAS)
	301	Research Institute for Microbial Diseases, Osaka Univeristy (RIMD)
	343	Culture Collection of Actinomycetes, Kaken Chemical Co. Ltd. (KCC)
	542	Institute for Microbial Resources (MR)
New Zealand	318	New Zealand Reference Culture Collection (NZRCC) (Dairy)
	335	New Zealand Reference Culture Collection (NZRCC) (Plant and Soil)
	376	New Zealand Reference Culture Collection (NZRCC) (Animal)
	457	New Zealand Reference Culture Collection (NZRCC) (Health)

Table V (continued)

Country	WDC collection no.	Service collection[a]
The Netherlands	133	Centraalbureau voor Schimmelculturen (CBS)
United Kingdom	140	Culture Centre of Algae and Protozoa (CCAP)
	154	National Collection of Type Cultures (NCTC)[b]
	238, 239	National Collection of Industrial and Marine Bacteria (NCIMB)[b]
	118	National Collection of Dairy Organisms (NCDO)
	126	National Collection of Plant Pathogenic Bacteria (NCPPB)
	214	Commonwealth Mycological Institute (CMI)[b]
	134	National Collection of Cultures of Wood-Rotting Macro-Fungi (FPRL)
	184	National Collection of Pathogenic Fungi (MRLL)
	169	National Collection of Yeast Cultures (NCYC)[b]
Union of Soviet Socialist Republics	342	USSR All-Union Collection of Microorganisms (VKM)
United States of America	1	American Type Culture Collection (ATCC)[b]

[a] Correspondents and addresses are given by McGowan and Skerman (1982).
[b] International Depository Authority for the deposition of patent cultures under the Budapest Treaty (1977).

intense project of short duration. These research collections are usually highly specialized and are often the only source of particular strains, serotypes, or genetic mutants. The value of these collections often lies in the strain variations represented. Such strains are indispensable in studies of variability within taxa. The continuity of these collections is vulnerable and often reaches a crisis when the research scientist retires or when the research project ends. At this time selected strains from these "orphan collections" must be accessioned into permanent collections. It is an advisable approach for research scientists to regularly

deposit representative strains on which they have published into service or institutional collections to avoid the loss of important strains at the termination of their career.

There are of course many collections which combine a number of the functions outlined above. Some institutional collections provide a service function as well by supplying cultures on demand to other institutions and research workers. Similarly many service collections provide cultures for research and identification services conducted in the institutions in which they are housed.

In addition to the primary function of collecting, preserving, distributing and exchanging cultures of microorganisms, culture collections act as reference centres which contribute towards and support studies of microorganisms and which promote their application. By their very nature culture collections are centres where the best methods of preservation of cultures are perfected and applied, where taxonomic and applied research is undertaken, and where microbiological methods are devised and standardized.

The importance of culture collections as a world resource has been established. One area where the importance of culture collections cannot be overstresssed is the contribution they make as an information resource. Culture collections maintain records on the characteristics of the cultures held, including valuable information on morphological, nutritional, genetic, physiological and other properties of interest. This type of information is frequently sought and the type of information required is becoming more detailed. The appearance of a new discovery published in the scientific literature often prompts a search for strains with similar properties.

In recent years there has been a universal recognition that the orderly development of all aspects of microbiology and biotechnology and their ability to contribute towards the best use of our world resources can be best served by a substantial data network. Such a network should encourage culture collections to collect and store information on their cultures and then exchange information by making it available to data centres, other collections and the microbiological community. Data storage and exchange has grown in importance as a culture collection function, due to the impetus at various national and international levels.

A number of large national service collections, such as the American Type Culture Collection and the Commonwealth Mycological Institute, have developed their own computer data bases and programmes for

information storage and retrieval, and in some developed countries such as Japan moves are under way to establish data bases at the national level (Komagata, 1977). Discussions are also under way to invesitgate the cooperative exchange of information between data centres at an international level.

It has been recognized that effective information exchange needs to be coordinated at the international level as well as at the regional and national level. It has also been recognized that an effective data network must include both developed and developing nations for their mutual benefit and development.

One of the major developments in information collection and exchange has come through the coordinated efforts of the United Nations Environment Program (UNEP), the United Nations Educational, Scientific and Cultural Organization (UNESCO), the International Cell Research Organization (ICRO) and the World Federation for Culture Collections (WFCC), which has resulted in the development of a network of Microbiological Resources Centres (MIRCENs) in developing and developed countries.

The MIRCEN concept was formulated at a UNEP meeting in Nairobi in 1974, and is aimed at establishing a regional network of collaborating institutions and laboratories with one serving as the coordinating centre for the region, and acting as a reference source and organizational centre for training courses, technological transfer and information exchange in the region (la Riviere, 1976).

The MIRCEN concept is now in an advanced stage of development with the establishment of three networks of institutions in the developing regions of Africa, Latin America and Asia with complementary laboratories in developed countries such as Australia, Sweden, the United Kingdom and the United States of America (Table VI).

The coordinating MIRCEN for information exchange is the World Data Centre for Microorganisms (WDC) in Brisbane. The establishment of the WDC resulted from the recognition that the increased demands for information on the location, history, and characteristics of strains in culture collections could be best met by the development of an adaptable system for storing, retrieving and exchanging information which could be used by all microbiologists (McGowan and Skerman, 1982). The WDC owes its continued existence to the financial support of UNEP, UNESCO and the University of Queensland. At the present stage of development the Centre is able to provide information on the

TABLE VI
The MIRCEN network[a]

Bangkok Fermentation, Food, and Waste Recycling MIRCEN
Thailand Institute of Scientific and Technology Research,
196 Phahonyothin Road,
Bangkhen,
Bangkok 9,
Thailand.

Beltsville Rhizobium MIRCEN
Cell Culture and Nitrogen Fixation Laboratory,
Room 116, Building 011-A,
BARC-West,
Beltsville,
Maryland 20705,
U.S.A.

Birmingham Biodeterioration MIRCEN
56 Peter's College,
University of Aston in Birmingham,
College Road,
Saltley,
Birmingham B8 3TE,
United Kingdom.

Brisbane World Data Centre MIRCEN
Department of Microbiology,
University of Queensland,
St. Lucia, 4067,
Brisbane,
Australia,

Cairo Biotechnology MIRCEN
Faculty of Agriculture,
Ain Shams University,
Shobra-Khaima,
Cairo,
Arab Republic of Egypt.

Dakar MIRCEN
Institut Sénégalais de Recherches Agricoles (Bambey),
c/o rue de Thiony x Valmy,
B.P. 3120,
Dakar,
Senegal.

Guatemala Biotechnology MIRCEN
Applied Research Division,

Central American Research Institute for Industry,
Ave. La Reforma 4-47 Zone 10,
Apartado Postal 1552,
Guatemala.

Nairobi Rhizobium MIRCEN
Department of Soil Science and Botany,
University of Nairobi,
P.O. Box 30197,
Nairobi,
Kenya.

Hawaii Rhizobium MIRCEN
NifTAL Project,
College of Tropical Agriculture and Human Resources,
University of Hawaii,
P.O. Box "0",
Paia,
Hawaii 96779,
U.S.A.

Porto Alegro Rhizobium MIRCEN
IPAGRO,
Caixa Postal 776,
90000 Porto Alegro,
Rio Grande do Sul,
Brazil.

Stockholm Biotechnology MIRCEN
Department of Bacteriology,
Karolinska Institute,
Fack,
S-104 01 Stockholm,
Sweden.

---

[a] Data from Rosswall and Da Silva (1982).

location of any particular species of microorganism in the 356 registered culture collections. The Centre is able to produce directories or
sub-directories for any collection or group of collections, country or
region; or for any particular organism of economic or other importance
(V.B.D. Skerman, personal communication). The Centre has recently
published the second edition of the *World Directory of Collections of
Cultures of Microorganisms* (McGowan and Skerman, 1982) and the
second edition of the *IBP World Catalogue of Rhizobium collections*
(Skinner, Hamatova and McGowan, 1983) is in press.

In 1975 the WDC embarked on a new phase of operation involving the acquisition of strain information of the type normally provided in the more detailed culture collection catalogues. After consultation with collaborating culture collections, the WDC introduced the standard accession data form (SCC−4 form) (Skerman and Leveritt, 1977). This form is shown in Figure 1 and a second form designed specifically for the genus *Rhizobium* is shown in Figure 2. The objective has been to encourage all collections to obtain and exchange the same level of basic information about strains, and to establish a common mode of accession which would improve intercommunication between collections themselves and between collections and the WDC. Receipt of strain data will allow the WDC to expand its services to much more specific information and to publish strain catalogues in addition to species directories.

Information exchange at the administrative and professional levels also occurs amongst culture collections, their personnel and their users. Culture collections receive requests not only for the cultures held in their own or other collections, but also for many other types of information. Information on quarantine, customs and postal regulations, identification, nomenclature, preservation and deposition of cultures for patent purpose is frequently sought. Culture collections should be able to advise on such matters. In order to facilitate the exchange of such information and foster the interests of culture collections, a number of international and national organizations have been formed. Principal amongst these are the World Federation for Culture Collections (WFCC), which is an Interdisciplinary Commission of the International Union of Biological Sciences, and a number of national federations and committees which have been established in Australia, Canada, Czechoslovakia, Japan, Korea, Latin America, New Zealand, the United Kingdom and the United States of America.

The way in which the WFCC is contributing towards the conservation of our microbial heritage is best seen from its stated objectives (Statutes, 1972):

- to establish an effective liaison between persons and organizations concerned with culture collections and between them and the users of cultures
- to encourage the study of procedures for the isolation, culture, characterization, conservation, and distribution of microorganisms
- to promote the training of personnel for the operation of culture collections

NOMENCLATURE DATA

Genus	Species, subspecies, infrasubspecies	Kind*
=100 Authors of the name (*with dates*) (e.g. *Rhizobium phaseoli* Dangeard 1926)		
=110 World Directory Collection Number	Collection Acronym	Collection Accession Number
SYNONYMS		

HISTORY

=120 Received as (*Genus*)	Species, subspecies, infrasubspecies	=125 Date received / / Day Month Year
=130 Received from (*Name and address of person or collection*) As		=140 Strain Designation
=150 Who received it from (*Name and address of person or collection*) As		=160 Strain Designation
=170 Who received it from (*Name and address of person or collection*) As		=180 Strain Designation
=185 Any previous history (*include accession number*)		
=190 Also held in following permanent collections (*include World Directory collection number, acronym and accession number e.g.* ATCC 12345)		

ORIGIN

=200 Isolated or derived from (*If plant, animal or protist give genus and species name and common name of the host*)		
		=210 Anatomical part (*if applicable*)
220 Original collection site		=221
		Country
=222	=223	=224
State/Province	Nearest town	Distance and direction from town
=225	=226	=227
Altitude	Latitude (*indicate North or South*)	Longitude (*indicate East or West*)
=230 Isolated by (*Name*)	=235 Strain Designation	=240 Date / / Day Month Year
=250 Identified by		=260 Date / / Day Month Year

Fig. 1. Standard SCC-4 accession data form (Reproduced with the permission of the World Data Centre for Microorganisms).

MAINTENANCE

=270 Method of Preservation (*Please tick whichever is applicable*) and =280 Temperature of Storage

As Original Source Material ___ °C   Agar Culture ___ °C

Lyophilised Culture ___ °C   Liquid Nitrogen ___ °C

In Soil ___ °C   Under Oil ___ °C

Dessicated on beads ___ °C   Other (*Specify*).................... ___ °C

=290 Preservation suspending medium

=300 Growth Medium (*name and reference, or give details on separate page*)   =310 Temperature

=320 Growth conditions (*e.g. aerobic, anaerobic, special gas mixture, light*)

=322 Incubation time before examination   =327 Period between subcultures

SPECIAL FEATURES AND USAGE (*e.g. for citric acid production, Type strain etc.*)

=330

REFERENCES (*Author, Journal, volume, pages, year*)

=340

=990

*Kind — B = bacteria, A = algae, F = fungus, Y = yeast, P = protozoa, AV = animal virus, BV = bacterial virus

PV = plant virus, IV = insect virus, L = lichen, TC = tissue cultures.

If space is inadequate to reply to any question, please submit additional information on separate pages, identified by the appropriate reference number (e.g. =310).

[3/77]

NOMENCLATURE DATA

Genus RHIZOBIUM	Species, subspecies, infrasubspecies	Type B
=100 Author(s) of the name *(with date(s))* *(e.g. Rhizobium phaseoli Dangeard 1926)*		
=110 World Directory Collection Number	Collection Acronym	Collection Accession Number

HISTORY

=120 Received as (Genus, species, subspecies, infrasubspecies)		=125 Date received / / Day Month Year
=130 Received from *(Name and address of person or collection)*	As	=140 Strain designation
=150 Who received it from *(Name and address of person or collection)*	As	=160 Strain designation
=170 Who received it from *(Name and address of person or collection)*	As	=180 Strain designation
=185 Any previous history (include accession number)		
=190 Also held in following permanent collections *(Include acronym and accession number, e.g. ATCC 12345)*		

ORIGIN

=200 Isolated from

=201 Genus _____   =202 Species _____   =203 Cultivar _____

=210 Part of plant *(Tick applicable block)*

☐ Root nodule   ☐ Other (specify) ....

=220 Original collection site	=221 _____ Country	
=222 _____ State/Province	=223 _____ Nearest town o , ,,	=224 _____ Distance and direction from town
=225 _____ Altitude	=226 _____ Latitude (indicate North or South)	=227 _____ o , ,, Longitude (indicate East or West)

Collection site soil properties

=350 Soil family _____   =358 Phosphorous (ppm) _____
=352 pH (water) _____   =360 P-extraction solution used _____
=354 pH (N KCl) _____   =362 Salinity (m moles/cm) _____
=356 pH (.01 M CaCl$_2$) _____   =364 Base saturation (%) _____

=230 Isolated by *(Name and address)*	As	= 235 Strain designation	=240 Date / / Day Month Year
=250 Identified by			=260 Date / / Day Month Year

Fig. 2. SCC-4 accession data form for the genus *Rhizobium* (Reproduced with the permission of the World Data Centre for Microorganisms).

MAINTENANCE

WFCC Form SCC-4(R)

=270 Method of preservation *(Please tick whichever is applicable)* and =280 Temperature of storage			
Original source material	°C	Agar culture	°C
Lyophilised culture	°C	Liquid nitrogen	°C
In soil	°C	Under oil	°C
Desiccated on beads	°C	Other *(Specify)* _____	°C

=290 Preservation suspending medium

=300 Growth medium *(Name and reference, or give details on separate page)*	=310 Temperature

=320 Growth conditions *(e.g. aerobic, anaerobic, special gas mixture, light)*

=322 Incubation time *(Tick applicable block)*	=327 Period between subcultures
Slow (> 5 days) Fast (< 5 days)	

SPECIAL FEATURES AND USAGE *(e.g. for citric acid production, Type strain etc.)*

=370 Other legumes nodulated	=373 Effectiveness (E, PE, I)	=376 Suitability as inoculant (Excellent, good, fair, poor)

Tolerance to stress: Indicate either (a) *range* or (b) *high, medium, low* tolerance

=380 As free-living organisms:

pH _____ , Al^{+++} ___ ___ , Mn^{++} ___ ___ , NaCl _____ , Temp. _____ °C

=390 As symbionts in host nodules:

pH _____ , Al^{+++} ___ ___ , Mn^{++} ___ ___ , NaCl _____ , Temp. _____ °C

=330 Other special features *(CHO metabolism, etc.)*	=335 Reaction in standard YMA
	Acidic Basic

REFERENCES *(Journal, volume, pages, year)*

=340

If space is inadequate to reply to any question, please submit additional information on separate pages identified by appropriate reference number *(e.g. =310).*

E = Effective   PE = Partially effective   I = Ineffective

- to promote the establishment of an international data service concerned with the location of, and information about, microorganisms maintained in culture collections and to publish a world directory of culture collections and list of species maintained therein
- to promote and aid the establishment of culture collections
- to promote the establishment of special reference collections and identification services and aid existing identification services
- to establish official means of communication
- to organize conferences and symposia on topics and problems of common interest
- to attempt solution of international problems of distribution of cultures of microorganisms that may arise through quarantine regulations, and
- to attempt to provide for the perpetuation of important collections or cultures

The WFCC has made major contributions in all these areas, and through its subcommittees advises members and other organizations on matters relating to patent depositions, postal and quarantine regulations, data coding, education, and the safeguard of endangered collections. Affiliated collections are also required to meet specified standards (Report, 1975).

Prominent amongst the activities of the WFCC is education and training. Training courses on culture collections and preservation technologies are often held concurrently with International Conferences on Culture Collections in developing countries. Many WFCC members are also involved in training courses sponsored by the UNEP/UNESCO/ICRO/MIRCEN network. Courses have been organized in Australia, Brazil, Czechoslovakia, Egypt, Ethiopia, Guatemala, Hong Kong, India, Indonesia, Japan, Kenya, Kuwait, Malaysia, Mexico, New Zealand, Nigeria, Philippines, the Republic of Korea, Singapore, Sri Lanka and Thailand. The courses cover such subjects as continuous culture systems, enzyme kinetics, microbial engineering, fermentation technology, conservation of genetic stocks for industrial efficiency and output, formulation of soil inoculants, development of waste recovery system, microbial environmental control and management, and other trends in applied microbiological research and development (Annon., 1979).

There are of course many other functions which culture collections may have because the scope of culture collection responsibilities to their

funding bodies is often as collectively diverse in nature as the organisms they hold. Such functions include the deposition of cultures for patent purposes, identification and typing services, and the distribution of cultures. These areas will be dealt with under the management of culture collections.

There is, however, one final function which should be dealt with here and that is the question of research in a culture collection. In my view it is highly desirable for culture collection staff to be involved in some aspect of research. Involvement in a research program helps to atune the staff to developments and trends in microbiology and to instil an awareness of the basic problems of culture isolation, maintenance, preservation and characterization. Further, research with or around the collection with which the staff are entrusted tends to alleviate the routine nature of some aspects of culture collection work, reduces the remoteness of the long-term objectives of culture collections, and promotes an immediate sense of usefulness for the cultures. This can only be good for culture collections and their staff.

THE MANAGEMENT OF CULTURE COLLECTIONS

The management of a culture collection involves a unique organizational system which may be divided into a number of distinct areas of operation. The procedures followed in the larger service collections and some institutional collections have been reviewed. [See for example Martin (1963), Clark and Loegering (1967), Iizuka and Hasegawa (1970), and Pestana de Castro *et al.* (1976)]. Considerable details on the policies, functions and management of the National Collection of Type Cultures and the National Collection of Industrial Bacteria have been reported by Lapage *et al.* (1970b). The latest catalogues of the American Type Culture Collection (1982) and the German Collection of Microorganisms (1983) also provide detailed information on the precedures adopted in those collections, as do the catalogues of most service collections. The procedures which have been successfully followed over more than twenty years in the diverse culture collection at the University of Queensland, have been described by Skerman (1976). The desired requirements for a good industrial culture collection have been reviewed by Hesseltine and Haynes (1973).

The management of a culture collection is conveniently summarized in the flow chart shown in Figure 3. The operations can be grouped into

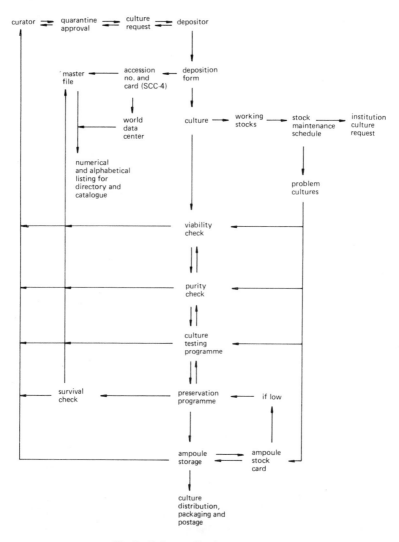

Fig. 3. Culture collection managment.

a number of areas each with particular problems but all of which interact in the overall management of the collection. These areas of operation are culture accessioning or deposition, documentation, culture maintenance, culture testing, culture preservation and the distribution of cultures.

The overall management of the collection is under the control of the curator who is responsible for implementing culture collection policy, the collection's programmes, and for the direction and training of staff. Curators of collections hold very responsible positions. They should be well qualified and motivated towards their work and need a sound knowledge of systematics and systematics procedures. The curator should oversee the acquisition of new cultures. In doing so he should be familiar with literature on new microorganisms and seek to obtain those which may have an immediate or future use in the research or service programmes of the collection. The curator should also keep abreast with trends in culture collections, relevant techniques and equipment, and matters relating to safety in culture collections (Darlow, 1969; Lapage *et al.*, 1970b; Ashby, 1975; Godber, 1975). It is essential that culture collections have adequately trained support staff and adequate modern facilities for the preservation and study of cultures and an adequate budget. In the words of Skerman (1978) "to provide less is to ask the impossible and is false economy".

The accessioning of new cultures into the collection may occur through acquisition from outside sources or may occur through a deliberate isolation programme. To take the case of industrial collections, suitable cultures may be selected from natural sources or obtained from other culture collections. Industrially important cultures in culture collections may, however, be restricted by patents either pending or granted. Isolates may have to undergo further genetic manipulation in laboratories to select desired characters before deposition in the collection. The procedures for the selection of industrially suitable strains from nature and the improvement of their performance have been detailed by Calam (1964) and Hesseltine and Haynes (1973). Many industrially important cultures are held in proprietary culture collections, service collections or private collections. In general, cultures in many proprietary collections are carefully guarded by industrial companies which often only release to service collections those cultures covered by patents in order to satisfy patent regulations.

The point has already been stressed on the importance of adequate documentation in culture collections. All collections design and use forms for recording the history and characteristics of cultures and for the generation of alphabetical and numerical lists and catalogues. Some of the forms used in major service collections have been published (see for example Lapage *et al.*, 1970 b; American Type Culture Collection, 1982).

In the collection in my department we have routinely used the World Data Centre's SCC–4 accession data forms (see Figure 1 and 2) and found them most suitable. Use of these forms also has the advantage of aiding the ease of information exchange, which has already been discussed, as have the reasons for seeking the detailed accession data. The more data available the more valuable the strain becomes. One point which should be given considerable attention is the assignment of the strain number. Care should be taken to ensure that the combination of number and collection acronym is unique so that no confusion arises in the literature.

For reasons of financial and labour savings and minimization of genetic change, most culture collections engage themselves in the minimal amount of stock culture maintenance. However, there are various stages in culture collection procedures where it is necessary to maintain cultures in an actively growing state. Typical cases are during initial deposition, pre-preservation testing and more detailed testing programmes. Also there are some cultures for which no reliable long-term preservation methods have been devised. The collection at the University of Queensland, for example, may at any one time be required to actively maintain some hundreds of cultures of a very diverse nature and diverse maintenance requirements in order to service the broad teaching and research programmes. A simplified procedure using punch cards for characters such as growth medium, period of subculture, incubation time, growth temperature and other growth environmental requirements, has been devised for the generation of stock maintenance schedules (Skerman, 1976). This facilitates the setting out of cultures in groups according to medium and growth conditions and minimizes continual time consuming referral to records. The task of setting up the schedule can be done annually and slight changes made as required during the year.

The amount of culture testing which is carried out in a culture collection varies considerable with the collection and the nature of the cultures. The depth of knowledge of some well studied strains will be considerably greater than others. Systematics orientated collections are often engaged in detailed testing of groups of strains in various taxa but for many industrially orientated collections this type of data is often not required. Industrial collections may collect cultures into broadly based taxonomic groups and place more emphasis on a particular metabolic or other property. But whatever the reason for the culture being in the

collection it should be remembered that culture testing plays a very important role in the quality control of the collection's preservation programmes. A culture preserved in an ampoule today may be required tomorrow, in a decade or not for another generation. It is imperative then that both pre- and post-preservation culture testing is carried out to ensure that the culture has retained the key characters for which the culture was preserved and that the culture is pure and viable. Strains selected for use in a biotechnological process usually have a unique set of characters and productive efficiency. Quality control testing is essential to ensure that these characters which have been selected or developed are preserved in as stable and reliable a state as possible.

The distribution of cultures is regulated by various local, national or international bodies and culture collections must build these constraints into their management procedures. In particular, microbiologists who use the local and international postal service should be thoroughly conversant with the postal and quarantine regulations governing the shipment of cultures. This applies particularly to culture collections who may distribute many cultures, some of which may pose a threat to human, animal or plant health, if not properly handled.

It would be hypocritical indeed to be singularly concerned with the conservation of our microbial heritage without due concern for the conservation of other life forms. Basically then, all microbiologists shipping cultures should do so in accordance with the current regulations and with a concern for the people who handle the package and the other mail should such a parcel be broken. One should also be concerned with the moral and ethical implications of dispatching cultures which may have a harmful effect on the health, economics or ecology of the recipient nation. For this reason it is wise not to send unsolicited cultures before ascertaining whether such cultures will be accepted. Failure to do so often causes considerable concern and effort for persons receiving such cultures.

The packaging requirements specified for perishable biological substances were agreed upon between the Universal Postal Union and the International Air Transport Association, and have been incorporated into the international legislation governing exchanges of mails between countries. The packaging requirements have been laid down for the categories of infectious and non-infectious substances. Packages must be properly packed, labelled and exchanged between recognized laboratories by registered mail only. Detailed packaging instructions are avail-

able (Lapage et al., 1970b; American Type Culture Collection, 1982; WHO Epidemiological Record, 1978). Some countries such as the United States of America also place export restrictions on the shipment of cultures (American Type Culture Collection, 1982).

The customs, quarantine and postal requirements are so varied and relatively unknown to the community at large that there is an urgent need for the compilation of an international guide to the shipment of cultures. In the meantime the greatest problem facting quarantine authorities is ignorance of, or contempt for, quarantine regulations, and we are all aware of breaches of these regulations. The shipment of cultures should be done openly in accordance with existing postal and quarantine regulations and with a sense of respect for our colleagues and mankind in general.

THE PRESERVATION OF MICROBIAL CULTURES

The objective of preservation methods is to maintain viability and genetic stability by reducing the organism's metabolic rate. A reduction in metabolic activity is usually achieved by withholding nutrients, water and oxygen, by reducing the storage temperature or by combinations of these.

The choice of method depends on the nature of the microorganism, the availability of equipment and skilled personal, and on the preservation objective. The method of preservation used also depends on whether the culture is to be preserved for a few days until a positive identification is made, for the duration of a research project or for future long-term reference. Within the context of microbial conservation the methods of choice should be directed towards long-term preservation wherever they are available. For this purpose most methods rely on keeping cultures in a frozen or a desiccated state, and while such procedures are suitable for many organisms, there are a significant number of organisms for which there is no reliable or satisfactory method of long-term preservation.

The preservation methods used in collections reflect the different biological properties of the various groups of microorganisms such as the bacteria, viruses, fungi, yeasts, algae and protozoa, and their responses to changes in their environment. Even within each group the successful preservation of cultures is subject to variability within genera

and species and even between strains within species. For this reason it is not possible for most collections to optimize the preservation for each culture. Rather, most collections apply generally accepted methods and in the event of these being unsuccessful it is necessary to investigate in more detail the correct conditions for the preservation of these more difficult cultures. The exception may be in industrial collections where more attention may be given to the reproducibility of the recovery rate and potency of cultures.

The application of the various methods of preservation to the different groups of microorganisms has recently been evaluated (Rogosa, 1981). In general the most succesful methods in terms of longevity and genetic stability employ freezing or desiccation. This assessment confirms current opinion that the methods of choice for long-term preservation are freeze-drying (lyophilization) and cryogenic storage. These methods may be applied across a wide variety of microorganisms. Because of strain variation it is unwise to place complete confidence in any one method and it is advisable to preserve cultures in more than one way as a matter of insurance against unpredicted loss.

The general methods which have been succesfully applied to the preservation of microbial cultures have been reviewed on a number of occasions in considerable detail (Nei, 1968; Lapage *et al.*, 1970b; Gherna, 1981; Sly, 1983). The particular methods in use in the National Collection of Industrial Bacteria have been described by Lapage *et al* . (1970b) and in the American Type Culture Collection by Hatt (1980). The problems associated with the preservation of cultures in industry have been discussed by Calam (1964) and by Hesseltine and Haynes (1973).

Traditionally, microbiologists have used periodic subculture to preserve cultures in the actively growing state and still do for periods during culture testing, or where alternative preservation facilities are not available, or have not been considered. The risk of selecting mutants during periodic subculture is high and the method is not recommended for the long-term preservation of cultures in culture collections if the culture can be preserved by more reliable methods. The period between subcultures can vary between one or two days and six or more months and depends on the nature of the organism, the composition and pH of the medium, the degree of aeration and dehydration, and the tempera-ture of storage. The conditions used for the maintenance of micro-

organisms is scattered through the research literature but some data has been compiled (see for example Lapage, Shelton and Mitchell, 1970; Gherna, 1981).

The most successful preservation methods rely on desiccation or freezing. Freeze-drying (lypophilization) has been widely used as the preferred method for long-term preservation in culture collections for many years. Freeze-drying is particularly suitable for collections which supply cultures on demand because of its broad applicability, economy of storage space and ease of distribution. Freeze-drying is the most technologically complex of the preservation methods in use, requiring the highest level of technical skill and high capital investment. The method is suitable for many types of microorganisms including most bacteria, yeasts, sporing-fungi and some viruses, but is generally unsuitable for non-sporing fungi, some viruses, algae and protozoa. Many cultures are able to survive for periods of twenty to forty years (Gherna, 1981; Rogosa, 1981), but there are others for which no suitable conditions have yet been devised.

The process of freeze-drying involves the removal of water vapour by vacuum sublimation from the frozen state thereby overcoming the problems associated with drying from the liquid state. For detailed descriptions of freeze-drying and procedural instructions see papers by Lapage *et al.* (1970b), Hatt (1980), Gherna (1981) and Sly (1983). There are two methods used for freezing the cell suspension prior to the drying process. In the pre-freezing method the cell suspensions are frozen in the ampoules before being dried under vacuum, freezing being achieved by mechanical freezer or by using a mixture such as dry-ice in ethanol. The alternative method is known as centrifugal freeze drying where the cell suspension is frozen by evaporative cooling under vacuum while the ampoules are spun in a low speed centrifuge to minimize foaming. Preservation media are essential for protecting cells from freezing damage and against overdrying. The choice of preservation medium depends on the organism and it must maintain the organism in a viable state and allow good recovery from the dried state. Preservation media usually contain high levels of serum, protein, carbohydrates or skim milk. Further information on the composition of preservation suspending media may be found in papers by Greaves (1964), Lapage *et al.* (1970a), Redway and Lapage (1974) and Hatt (1980).

Although slow uncontrolled dehydration from the liquid state is usually harmful, some strains of microorganisms which are sensitive to freeze-drying are able to be preserved by drying from the liquid state rather than the frozen state. The method was developed by Annear (1954, 1956, 1962) and has been successfully used to preserve bacteria, yeasts, fungi and viruses, A modified method developed by Banno *et al*. (1979) has been routinely used with success for bacteria and yeasts in a culture collection.

Most microorganisms including bacteria, yeasts, fungi, viruses, bacteriophages and some algae and protozoa can survive long-term storage in the frozen state by markerly reducing their metabolic rate. Animal, human and plant cells may also be successfully preserved by freezing. Microorganisms have been stored in freezers at temperature around −20°C and −70°C. The lower the temperature the less is the loss of viability and the longer the survival time.

The use of cryogenic storage by freezing at ultra-low temperature obtained by freezing in liquid nitrogen at −196°C has provided culture collections with a simple standardized technique for the preservation of a wide range of microorganisms and mammalian cells (Ashwood-Smith and Farrant, 1980; Bridges, 1966; Nei, 1968). Cryogenic storage offers a much reduced viability loss, a high degree of genetic stability and expected survival periods of greater than thirty years (Gherna, 1981; Rogosa, 1981). Detailed procedures followed in culture collections may be obtained from papers by Hatt (1980), Gherna (1981) and Sly (1983). The initial cost of cryogenic equipment is moderate on a small scale but there is a continuing outlay for the replenishment of liquid nitrogen for which a guaranteed regular supply is essential. The reliability of cryogenic storage justifies its cost.

A variety of cryoprotective agents and freezing protocols are needed to successfully preserve the large range of microorganisms (see for example Hatt, 1980). In general glycerol or dimethyl-sulphoxide (DMSO) at concentrations at 5−10% afford protection but in some cases higher concentrations or other cryoprotectants such as methanol (Morris and Canning, 1978; Hatt, 1980) are required. The rate of cooling should be slow and controlled down to −20°C to −40°C and then rapid to the final freezing temperature. The rate of thawing should be as rapid as possible (Goos, Davis and Butterfield, 1967). Rapid freezing rates and absence of cryoprotectant may lead to intracellular

ice crystal formation and electrolyte imbalance and cause lethal cell damage. In general, bacteria, yeasts and fungi are less sensitive to freezing damage than algae, protozoa and tissue cultures where the choices of cryoprotectant and freezing rate are very significant factors. The significant increase in the numbers of algae and protozoa in culture collections during the last ten years has in the main part been due to the development of successful cryogenic techniques resulting from considerable painstaking research (see for example Morris and Clarke, 1976; Morris and Canning, 1978).

However, not all laboratories may have facilities for freeze-drying or cryogenic storage. Culture collections in developing countries may not have ready access to the necessary equipment, trained staff, or an economical and reliable supply of liquid nitrogen. In this case there are a number of simple but well proven methods of preservation which are worthy of consideration. They require little capital expenditure and may be suitable for laboratories with limited resources. Although these methods may not be suitable across as broad a range of cultures as freeze drying and cryogenic storage, and are more suitable for short and medium-term preservation, they are considerably better than routine subculture.

Some example of alternative methods which have been used include storage under mineral oil, preservation in sterile soil (Jensen, 1961), storage in sterile distilled water (De Vay and Schnathorst, 1963; McGinnis *et al.*, 1974), preservation on porcelain beads (Norris, 1963), preservation on silica gel (Perkins, 1962; Leben and Sleesman, 1982), preservation in gelation discs (Stamp, 1947), and preservation over phosphorus pentoxide in vacuo (Soriano, 1970). Further procedural details and their application in culture collections may be obtained by referring to papers by Lapage *et al* . (1970b), Gherna (1981) and Sly (1983).

FUTURE TRENDS

Culture collections face considerable challenges in the future in most phases of their function and operation. In the past, culture collections have been developed to cater for the requirements of all aspects of microbiology as the science developed. There has been a continual evolution of interests to meet the demands of education, research, medical, veterinary and agricultural microbiology, and applied micro-

biology and biotechnology. The early development of industrial culture collections was concerned with fermentations, food spoilage and pre-servation, agriculture, public health and vaccine production. The next major development in culture collection activity in industry occurred during the "antibiotic era" when culture collections were concerned with the isolation of enormous numbers of environmental isolates and mutants capable of producing novel substances.

There has been a realization that microbial genetic resources may be exploited even more for the management of the environment and for the benefit of mankind. There is considerable interest in the use of micro-organisms for the detoxification and bioconversion of wastes, the purifi-cation of polluted waters, the controlled fermentation of indigenous foods, the microbial fixation of nitrogen, the production of methane fuel from garbage and manure, and the conversion of solar energy by microbial photosynthesis (Anon., 1979). All these areas of endeavour will require the selection of new environmental isolates and the screening of existing gene pools for organisms with suitable capabilities.

It appears that the next wave of industrial culture collections will be concerned with the "biotechnology era" using genetically engineered mutant strains and hybridomas. These cultures will pose new challenges for their successful preservation. Genetically engineered strains contain foreign extrachromosonal DNA and some are genetically unstable. Considerable research on optimal preservation methods for these strains will be required.

The move towards quality control of media and diagnostic and analytical procedures will mean even greater dependence on the supply of control cultures by culture collections. The American Type Culture Collection and a number of supply companies now provide kits of cul-tures for this purpose. No doubt the cultures available will expand to meet future demands. Innovation in the preservation and presentation of cultures in kit form for quality control and comparative purposes will need to keep pace with developments in rapid and automated methods in microbiology.

It is likely that more culture collections will find themselves involved in the deposition and supply of cultures in connection with patent applications. The reasons for this are twofold. Firstly there is an increasing use of microorganisms for novel purposes and greater aware-ness of the commercial benefit acruing by patent protection. Secondly, the adoption of the Budapest Treaty on the International Recognition of

the Deposit of Microorganisms for the Purpose of Patent Procedure in 1977 opened up the way for certain culture collections to acquire status as an International Depository Authority (IDA) for this purpose. Such status places obligations on a collection to maintain prescribed standards and to follow prescribed procedures for the deposition and release of patent cultures. But more importantly recognition of the collection as an IDA requires government assurances regarding its continuous existence and virtually guarantees the permanency of the collection.

In recent years there has been a developing realization that the preservation of cultures in culture collections is in fact an act of conservation. To some this may seem an unusual concept when compared with the conservation of the larger life forms such as animals and plants. The justification for this concept lies in the fact that, because of the vastness of the environment and the continuing evolution of characteristics, it is often extremely difficult to reisolate exactly similar strains of microorganisms from nature. The loss of a culture may in effect be as final as the extinction of an animal or plant — a matter which would cause considerable international concern and indignation. The realization has been that it is far more economical and secure to conserve our microbial cultures in culture collections than to face the uncertainty of attempting to reisolate cultures with desired characteristics at prohibitive cost.

Culture collections are entrusted with the conservation of an invaluable resource which is part of our natural world heritage. The significance of the irreplaceable gene pool represented in culture collections may only be fully appreciated in the light of new scientific discoveries and technological developments. Culture collections then are deserving of more recognition than they may have had in the past and their support is indeed worthy of considerable consideration in the future development of applied microbiology and biotechnology.

REFERENCES

American Type Culture Collection (1982). *Catalogue of strains* I, 15th ed. American Type Culture Collection, Rockville, Md.
Annear, D. I. (1954). 'Preservation of Bacteria.' *Nature* **174**, 359.
Annear, D. I. (1956). 'The Preservation of Bacteria by Drying in Peptone Plugs.' *Journal of Hygiene* **54**, 487.
Annear. D. I. (1962). 'Recoveries of Bacteria After Drying on Cellulose Fibres (A

Method for the Routine Preservation of Bacteria).' *Aust. J. Exp. Biol. Med. Sc.* **40**, 1—8.

Anon. (1979). 'A MIRCENs network to safeguard nature's invisible assets.' IUBS Newsletter **15**, 21—24.

Ashby, Lord. (Chairman) (1975). *Report of the Working Party on the Experimental Manipulation of the Genetic Composition of Microorganisms.* Her Majesty's Stationary Office, London.

Ashwood-Smith, M. J. and Farrant, J. (1980). *Low Temperature Preservation in Medicine and Biology.* Pitman Medical Ltd., Tunbridge Wells, U.K.

Banno, I., Mikata, K. and Sakane, T. (1979). 'Viability of Various Yeasts after L-drying.' *IFO Res. Comm.* **9**, 27—34.

Banno, I. and Sakane, T. (1979). 'Viability of Various Bacteria after L-drying.' *IFO Res. Comm.* **9**, 35—45.

Bridges, B. A. (1966). 'Preservation of Microorganisms at Low Temperature.' *Lab. Pract.* **15**, 418.

Budapest Treaty (1977). *International Recognition of the Deposit of Microorganisms for the Purpose of Patent Procedure.* World Intellectual Property Organization, Geneva.

Calam, C. T. (1964). 'The Selection, Improvement, and Preservation of Microorganisms.' *Prog. Indust. Microbiology* **5**, 1—53.

Clark, W. A. and Loegering, W. Q. (1967). Functions and Maintenance of a Type-Culture Collection.' *Ann. Rev. Phytopathol.* **5**, 319—342.

Darlow, H. M. (1969). 'Safety in the Microbiology Laboratory.' In *Methods in Microbiology* (J. R. Norris and D. W. Ribbons eds.) Vol. 1, Ch. VI, pp. 169—204. Academic Press, London.

De Vay, J. E. and Schnathorst, W. C. (1963). 'Single-Cell Isolation and Preservation of Bacterial Cultures'. *Nature* **199**, 775—777.

Fernandes, F. and Costa Pereira, R. (eds.) (1977). *Proceedings of the Third International Conference on Culture Collections.* University of Bombay, Bombay.

German Collection of Microorganisms (1983). *Catalogue of Strains,* 3rd edition. Deutsche Sammlung von Mikroorganismen, Göttingen.

Gherna, R. L. (1981). 'Preservation.' in *Manual of Methods for General Bacteriology* (P. Gerhardt, R. G. E. Murray, R. N. Costilow, E. W. Nester, W. A. Wood, N. R. Krieg and G. B. Phillips, eds.), Ch. 12, pp. 208—217. American Society for Microbiology, Washington.

Godber, G. (Chairman) (1975). 'Report of Working party on the Laboratory Use of Dangerous Pathogens.' *Nature* **255**, 362.

Goos, R. D., Davis, E. E. and Butterfield, W. (1967). 'Effect of Warming Rates on the Viability of Frozen Fungous Spores.' *Mycologia* **59**, 58—66.

Greaves, R. I. N. (1964). 'Fundamental Aspects of Freeze-drying Bacteria and Living Cells.' In *Aspects Theoretiques et Industriels de la Lyophilisation.* (L. Rey, ed.), pp. 407—410, Hermann, Paris.

Hatt, H. (Ed.) (1980). *American Type Culture Collection Methods. I. Laboratory Manual on Preservation, Freezing and Freeze-Drying.* American Type Culture Collection, Rockville.

Hesseltine, C. W. and Haynes, W. C. (1973). 'Sources and Management of Microorganisms for the development of a Fermentation Industry. *Prog. Indust. Microbiology* **12**, 1—46.

Iizuka, H. and Hasegawa, T. (ed.) (1970). *Proceedings of the International Conference on Culture Collections*. University Park Press, Baltimore.

Jensen, H. L. (1961). 'The Viability of Lucern Rhizobia in Soil Culture.' *Nature,* **192,** 682.

Komagata, K. (1977). 'The Japan Federation for Culture Collections and Plans of a Data Center for Microorganisms'. In *The Proceedings of the Fifth Biennial International CODATA Conference* (B. Dreyfus, ed.) pp. 311—312. Pergamon Press, Oxford.

Lapage, S. P., Shelton, J. E. and Mitchell, T. G. (1970a). 'Media for the Maintenance and Preservation of Bacteria.' In *Methods in Microbiology* (J. R. Norris and D. W. Ribbons, eds.) Vol. 3A, pp. 2—133. Academic Press,London.

Lapage, S. P., Shelton, J. E., Mitchell, T. G. and Mackenzie, A. R. (1970b). Culture Collections and Preservation of Bacteria. In *Methods in Microbiology* (J. R. Norris and D. W. Ribbons, ed.), Vol. 3A, pp. 135—227. Academic Press, London.

La Riviere, J. W. M. (1976). 'The role of Microbiology in the work of International Organizations.' In *Souvenir Book,* Third International Conference on Culture Collections, Bombay.

Leben, C. and Sleesman, J. P. (1982). 'Preservation of Plant Pathogenic Bacteria on Silica Gel'. *Plant Diseases* **66**, 327.

Martin, S.M. (ed.) (1963). *Culture collections: Perspectives and Problems. Proceedings of the First International Specialists' Conference on Culture Collections.* University Toronto Press, Toronto.

Martin, S. M. (1976). 'Regional Culture Collections in the Developing World.' In *Proceedings of the Second International Conference on Culture Collections* (A. F. Pestana de Castro, E. J. Da Silva, V. B. D. Skerman and W. W. Leveritt eds.) UNESCO/UNEP/ICRO/WFCC/World Data Centre for Microorganisms, Brisbane.

Martin, S. M. and Skerman, V. B. D. (1972). *World Directory of Collections of Cultures of Microorganisms.* Wiley-Interscience, New York.

McGinnis, M. R., Padhye, A. A. and Ajello, L. (1974). 'Storage of Stock Cultures of Filamentous Fungi, Yeasts, and Some Aerobic Actinomycetes in Sterile Distilled Water. *Applied Microbiology* **28**: 218—222.

McGowan, V. and Skerman, V. B. D. (1982). *World Directory of Collections of Cultures of Microorganisms,* 2nd ed. World Data Centre for Microorganisms, Brisbane.

Morris, G. J. and Canning, C. E. (1978). 'The Cryopreservation of *Euglena gracilis.' Journal of General Microbiology* **108**, 27—31.

Morris, G. J. and Clark, K. J. (1976). 'Cryopreservation of *Chlorella.' Les Colloques de l'Institute Nationale de la Sante et de la Recherche Medicale* **62**, 361—366.

Nei, Tokio (1968). 'Freezing and Drying of Microorganisms'. In *Papers Presented at the Conference on Mechanisms of Cellular Injury by Freezing and Drying in Microorganisms.* University of Tokyo Press, Tokyo.

Norris, D. O. (1963). 'A Porcelain Bead Method for Storing *Rhizobium.' Empire J. Exp. Agric.* **31**, 255—258.

Perkins, D. D. (1962). 'Preservation of *Neurospora* Stock Cultures with Anhydrous Silica Gel.' *Can. J. Microbiol.* **8**, 591—594.

Pestana de Castro, A. F., Da Silva, E. J., Skerman, V. B. D. and Leveritt, W. W. (eds.) (1976). *Proceedings of the Second International Conference on Culture Collections.* UNESCO/UNEP/ICRO/WFCC/World Data Centre for Microorganisms, Brisbane.

Porter, J. R. (1976). 'The World View of Culture Collections.' In *American Type Culture Collection 50th Anniversary Symposium, The Role of Culture Collections in the Era of Molecular Biology* (R. R. Colwell, ed.) p. 62–72. American Society for Microbiology: Washington.

Redway, K. F. and Lapage, S. P. (1974). 'Effect of Carbohydrates and Related Compounds on the Long-Term Preservation of Freeze-Dried Bacteria. *Cryobiology* **11**, 73–79.

Report (1975). 'Report of the World Federation for Culture Collections.' *Int. J. Syst. Bacteriol.* **25**, 90–94.

Rogosa, M. (ed.) (1981). *National Work Conference on Microbial Collections of Major Importance to Agriculture.* American Phytopathological Society, St. Paul, Mn.

Rosswall, T. and Da Silva, E. J. (ed.) (1982). *MIRCEN News*, No. 4. UNESCO, Paris.

Skerman, V. B. D. (1976). 'The Organization of a Small General Culture Collection.' In *Proceedings of the Second International Conference on Culture Collections* (A. F. Pestana de Castro, E. J. Da Silva, V. B. D. Skerman and W. W. Leveritt, eds.) pp. 20–40. UNESCO/UNEP/ICRO/WFCC/World Data Centre for Microorganisms, Brisbane.

Skerman, V. B. D. (1978). 'Importance of Culture Collections.' *Kükem Dergisi* **1**, 5–8.

Skerman, V. B. D. and Leveritt, W. W. (1977). 'Experience in Collection and Publication of Data in Microbiology.' In *The Proceedings of the Fifth Biennial International CODATA Conference* (B. Dreyfus, ed.) pp. 317–322. Pergamon Press, Oxford.

Skinner, F. A., Hamatova, E. and McGowan, V. F. (1983). *IBP World Catalogue of Rhizobium collections*, 2nd edition, (V. B. D. Skerman, ed.) World Data Centre, Bribane.

Soriano, S. (1970). 'Sordelli's Method for Preservation of Microbial Cultures by Desiccation in Vacuum.' In *Proceedings of the First International Conference on Culture Collections*, (H. Iizuka and T. Hasegawa, eds.) p. 269. University of Tokyo Press, Tokyo.

Sly, L. I. (1983). 'Preservation of Microbial Cultures. In *Plant Bacterial Diseases : A Diagnostic Guide.* (P. C. Fahy and G. J. Persley, eds.), Ch. 13, pp. 275–298. Academic Press, Sydney.

Stamp, L. (1947). 'The Preservation of Bacteria by Drying.' *J. Gen. Microbiol.* **1**, 251–265.

Statutes (1972). 'World Federation for Culture Collections Statutes.' *Int. J. Syst. Bacteriol.* **22**, 407–408.

WHO Eqidemiological Record (1978). 'Improved Procedures for the International Transport of Infectious Substances.' *Epidemiological Record No. 9.*

Culture Collection,
Department of Microbiology,
University of Queensland,
Brisbane,
Australia.

H. W. Doelle

Microbial Process Development in Biotechnology

INTRODUCTION

The development of a microbial process for the formation of biomass or products is aimed at maximizing three factors:

- the yield of product per gram of substrate
- the concentration of the product
- the rate of product formation.

In order to achieve this, the following main features of a microbial process development have to be observed:

(a) isolation, identification and initial selection of strains of organisms;
(b) determination of optimum values of nutritional requirements, temperature, pH and oxygen supply;
(c) modification of the genetic structure of the organism to increase the product formation;
(d) cell cultivation systems.

All four aspects are basically concerned with the adjustment of metabolic regulation in the organism, whereby metabolism means that all of the available carbon source is converted into biomass and the endproduct(s) of energy metabolism. Microbial process development can therefore be regarded as the ideal example for basic scientific research with an applied goal. The knowledge gained in such process development can then be translated into microbial process technology (Rogers, 1978), which in turn can be classified into high, intermediate, and low or village technology. Over the past decade, biotechnology has emphasized the development of technologies for organisms preserved in culture collections, which have never been investigated along the lines mentioned above (Doelle, 1982). If one wants to develop a technology of a process, one has to know the catalyst first. The latter, of course, is the appropriate microbe in question and the suitability for a process falls under microbial process development.

38

In terms of total biomas of our planet, microorganisms are equal to the animal kingdom (including human beings), together taking about half and higher plants the other half (Porter, 1981). The question was thus raised whether mankind has taken or is taking full advantage of this almost untapped natural resource. Microorganisms are still most frequently referred to as the cause of disease in human beings, animals or plants, and only slowly do we recognize that many more types are beneficial than harmful to higher forms of life. The reasons for this increasing awareness over the last decade is the realization that biological systems may be utilized for many new purposes in addition to food production (Chatel, 1980). It is the biological sciences which are expected to provide important potentialities for development in the second half of the twentieth century.

ISOLATION, IDENTIFICATION AND INITIAL SELECTION OF STRAINS OF MICROORGANISMS

A great number of culture collections, together with the recently established MIRCENs (Microbiological Resource Centres), contain large lists of microbial strains of more or less known characteristics (Sly, this volume). If one looks for a particular strain, the World Data Centre on microorganisms is available to locate the strain in the particular affiliated culture collections (McGowan and Skerman, 1982). The majority of these available strains, however, have neither been isolated nor explored with an aim for process development. It is therefore necessary to search for new, more suitable cultures, which possess the properties for producing the desired product in high yield (Riviere, 1977), or reinvestigate the existing strains from culture collections with the same aim, and at the same time economically utilize the available substrate. New cultures may be found by chance observations (e.g. Fleming's culture of *Penicillium notatum*) or more likely by a systematic search.

A systematic search for new cultures may depend on two major approaches: (1) the pure scientific and (2) the process development oriented search. Whichever direction is chosen, it is absolutely necessary to be well acquainted with the microorganisms, that is, one must be able to place them correctly into the system of living entities (Kockova-Kratochvilova 1981). Every isolation is connected with an evaluation of various features of microorganisms. The initial features in microbial

process development would undoubtedly be related to resource utilization (Campoz-Lopez, 1980) and/or product formation (Rose 1981). In sharp contrast to the usual requirements of academic research, organism isolation and initial selection for an industrial process is dependent on a range of criteria that are relevant to the optimization of the particular process (Bull *et al.*, 1979). There features may be morphological, physiological, genetic, immunological, etc., and the sum of all these features of a microorganism is referred to as its phenotype. A phenotype therefore represents any measurable characteristic or distinctive trait possessed by an organism. In contrast, the genotype represents all genes possessed by an organism. The genotype can be explored via the phenotypic expression.

The isolation, identification and initial selection of organisms for microbial process development depends therefore on the phenotypic expression of the organism. Despite the selective aim, one should not forget that every microbial culture must possess certain general attributes (Nakayama, 1981):

(a) the strain should be a pure culture and free of phages;
(b) the strain must be genetically stable;
(c) the strain must produce readily many vegetative cells, spores or other reproductive units;
(d) the strain should grow vigorously and rapidly after inoculation;
(e) the strain should produce the required product within a short period of time;
(f) if possible, the strain should be able to protect itself against contamination;
(g) the strain should produce the desired product, which should be easily separable from all others; and
(h) the strain should be amenable to change by certain mutagenic agents.

In most cases it is useful to isolate a culture from a natural source or decomposing or organic materials. Rapid screening techniques for testing the phenotypic expression normally combines isolation and selection simultaneously. The techniques used for these tests are numerous (Nakayama 1981) and depend, of course, on the expected phenotypic expression. Any isolated culture should immediately be deposited with a culture collection for maintenance and preservation (Dietz, 1982).

The isolation and identification of a new culture on phenotypic expression also gives some indication on the metabolism of the organism. It is of utmost importance, however, to investigate in details the basic metabolic processes (Doelle, 1981) of the organism as part of the selection programme. Traditionally, screening procedures are based on agar plate techniques or enrichment cultures. It should be realized that both methods can be very restrictive if one aims at certain microbial process development. The agar plate techniques are very important for enzyme- and antibiotic-producing strains. They give excellent results for polymer degradation (e.g. starch, cellulose, etc) by exoenzymes or antibiotic production, that is, phenotypic expression related to products excreted out of the cell. They also could be indicators for acidic or alkaline product formation. However, these procedures are very labour-intensive and time-consuming. Enrichment cultures, on the other hand, are carried out under substrate-excess conditions and thus select organisms on the basis of maximum specific growth rate. This characteristic may not be the key criterion for the process being developed. It also must be realized that in batch enrichment the time of sampling is important for the selection of the most desirable organism, since the growth conditions change as a function of time. It could therefore be possible to miss the particular stage when the particular organism is present in sufficient numbers to guarantee its isolation.

An attractive alternative has been developed during the last decade, which involves a continuous-flow enrichment technique (Veldkamp, 1970). This technique allows the selection and isolation of organisms on the basis of their substrate affinity (using a chemostat), maximum specific growth rate (using a turbido-stat), resistance to toxic materials, etc.

Different screening techniques select therefore different types of organisms and it is in the experimenter's hand to choose which one of these techniques would lead to the isolation and selection of the microbe wanted for the envisaged process development.

It was stated earlier that sound knowledge in microbial biochemistry, that is the basic metabolic processes, is an absolute requirement for a successful and speedy isolation and selection programme. Aerobic, facultative anaerobic and anaerobic organisms can be isolated selectively for their substrate specificity, growth rate or product formed.

DETERMINATION OF OPTIMAL NUTRITIONAL REQUIREMENT,
BIOMASS CONCENTRATION, TEMPERATURE, PH AND OXYGEN
SUPPLY

Despite their constant genotype, microbes are amazingly flexible in their
ability to alter their composition and metabolism in response to environmental change. By virtue of metabolic regulatory mechanisms, microbial
cells do not generally oversynthesize metabolites despite environmental
variations. Both, microbial growth and product formation therefore
occur in response to the environment. It is essential to understand the
relationship between the chemical and physical environment and regulation of microbial metabolism. The next step with our new culture is thus
concerned with the establishment of a medium economically usable on a
process scale. This goal automatically excludes solid media in favour of
liquid cultures, because the liquid media and cultures are amenable to
standard chemical engineering techniques and equipment. The requisite
conditions for growth of biomass in a culture are (a) a viable culture, (b)
an energy source, (c) nutrients to provide the essential material from
which the cell is synthesized, (d) the absence of inhibitors, and (e)
suitable physicochemical conditions (Pirt, 1975).

All microorganisms used for microbial process development require
organic compounds both as a source of carbon and energy. The element
carbon is the most abundant element and represents approximately 50%
of the biomass. In the case of algae and photosynthetic bacteria, the
energy source is light and the carbon source is carbon dioxide, chemo-
autotrophic bacteria can utilize inorganic compounds as energy source
and carbon dioxide as carbon source, whereas chemoheterotrophic
organisms require organic compounds for both. For any microbial
process development, the carbon source is therefore the largest
ingredient. If it is a limiting factor, the total biomass X is proportional to
the initial concentration of the organic source of carbon, which gives the
yield constant for the substrate and organism:

$$Y_s = X$$

This yield constant comes from the original definition by Monod:

$$\frac{dX}{dt} = - Y \frac{dS}{dt}$$

It is the carbon source, which therefore is predominant and is selected,
of course, from the substrate available. The intimate relationship

between the substrate as carbon and energy source can be found in cases when the energy yield (ATP) is known. In this situation there exists a proportionality between the number of moles ATP formed and the biomass produced. In anaerobic cultures, this yield factor is around 10 (Riviere, 1977). Under aerobic conditions, however, this yield factor varies greatly and is many-fold higher, since much larger proportions of carbon substrates are converted into biomass. It is possible to calculate the minimum quantity of a carbon substrate to obtain a specific yield of biomass. If one assumes a 50% biomass carbon requirement and required 50g of bacterial cells (dry weight) per litre, one would require

$$50 \times \frac{100}{50} \times \frac{50}{100} = 50 \text{ g carbon/litre}$$

The nutrients that are amenable to measurement and to cell growth are, apart from the carbon source, nitrogen source, oxygen, mineral salts and some specific growth factors, e.g. amino acids or vitamins (Cooney, 1981). The reliability of the method depends on the ratio of cell mass produced per unit nutrient consumed (cell yield), the accuracy of the analytical methods and the presence of substances which interfere with the analysis. Further on, if the substrate is also used for product formation and the ratio of product to cell mass is large and/or variable, a substantial error will result unless another independent measurement for product concentration is available to correct the measurement of substrate used for cell biosynthesis.

A very significant difference in approach for nutrient optimization depends on the kind of process envisaged: anaerobic or aerobic. In the latter case, oxygen is a vital nutrient, whereas redox potential replaces oxygen in anaerobic cultures. Optimization of anaerobic cultures requires a carbon balance between carbon substrate, biomass and product, whereas in aerobic cultures it is mainly a balance between carbon substrate and biomass. If all the requirements for growth are satisfied, then during an infinitely small time interval (dt) one expects the increase in biomass (dX) to be proportional to the amount (X) present and to the time interval

$$dX/dt = \mu X$$

whereby the differential coefficient (dX/dt) expresses the population growth rate and the parameter μ represents the rate of growth per unit amount of biomass ($1/X \; dX/dt$) or specific growth rate (Pirt, 1975;

Doelle, 1975; Cooney, 1981). The specific growth rate is the basic measure of the growth rate of any organism and should be used as an indicator for optimization of biomass production.

For anaerobic process development, the formation of the appropriate endproducts is often related to the specific growth rate and cell mass

$$\frac{dP}{dt} = \alpha\mu X + \beta X$$

where α and β are growth and non-growth associated constants. If one re-arranges this equation, one obtains the specific rate of product formation

$$q_P = \frac{1}{X} \frac{dP}{dt} = \alpha\mu + \beta$$

The correlation of product with growth depends on the relative value of α and β. The expression is very often expressed as $Y_{P/X}$, thus in purely growth-associated producted formation processes,

$$q_P = Y_{P/X}\mu$$

From these considerations it should become obvious that growth and product formation are bioconversion processes, in which the chemical nutrients fed to the growth vessel are converted to cell mass and metabolites. In using the expressions $Y_{X/S}, Y_{P/S}$, μ and q_p, it is possible to optimize the nutritional requirements for a microbial process in batch cultivation.

The microbial growth rate, however, is also a function of temperature, which has been described by the Arrhenius equation

$$\mu = Ae^{-Ea/RT}$$

where A is the Arrhenius constant, Ea is the activation energy, R is the gas constant and T is the absolute temperature. This means, of course, that temperature also affects the efficiency of the carbon-energy substrate conversions to cell mass and thus a variety of metabolic processes in the cell.

The most important factor in the optimization of a microbial process is therefore the design of the growth and production medium. From above outlined considerations, a first approximation of the minimum requirement could be achieved using the stoichiometry for growth and product formation. In contrast to academic research, several economic

and technical constraints should be considered or built in for microbial process development. These constraints include (Cooney, 1981) cost, availability of raw materials, requirements for specific carbon or nitrogen sources, recovery and pollution control. It should always be remembered that the ultimate goal or objective function is the development of a process with minimum cost per unit product.

Throughout this article it was mentioned that microbial cells have two main commercial applications. The first is a source of protein, primarily for animal feed. Since growth of a microorganism can be followed relatively easily, optimization methods have been developed and can be further explored along the lines mentioned earlier. A great number of industrial processes have been developed over the last decade and are outlined by Ringpfeil and Heinritz (this volume). The second commerical application is, however, the more difficult one, as it uses the microbial cell to carry out biological conversions and thus leads to organic chemicals (Eveleigh, 1981: Aharonowitz and Cohen, 1981) or enzyme production (Wang *et al.*, 1979). These biological conversions or microbial transformations can be accomplished with growing cells, non-growing cells, spores or even dried cells (Demain and Solomon, 1981). This area of microbial process development is still in its infancy, but develops at a staggering rate and almost certainly will have the biggest future. There are signs visible already for replacing certain chemical industries because of their heavy energy input demands. Biological conversions have many advantages over chemical conversions. Besides a strong energy input, chemical conversions require generally solvents and inorganic catalysts, both of which could be strong pollutants. They also may produce a number of unwanted by-products that must be removed. In contrast, biological conversions are carried out with water as solvent and at moderate biological temperatures. The commercially important products of microorganisms can be categorized into 3 groups:

(1) the large molecules such as enzymes
(2) the primary metabolic products (compounds essential for growth)
(3) the secondary metabolic products (compounds not required for growth).

In order to develop a microbial process in any of these three categories one should be aware that it is the DNA of the individual organism

or microbial cell that dictates the detailed synthesis of the enzymatic machinery. Although microbes are amazingly flexible in their ability to alter their composition and metabolism in response to environmental changes, the genetic make-up of the cell is not changed, but only the phenotypic expression.

It was mentioned earlier that whatever the changes in phenotypic expression may be, microbial cells do not generally oversynthesize metabolites whatever the environmental changes may be. The reasons for this behaviour are manifested in the genotype by virtue of metabolic regulatory mechanisms. It is of vital importance to know the coordination of these control mechanisms before any microbial process development is to commence.

Of the thousands of enzymes a cell is capable of producing according to its genetic code, some are always present (constitutive), whereas others require their substrate or analogous of these to be present (inductive). Induction is necessary in order to avoid wastage of energy or amino acids in making unnecessary enzymes; but that when needed, these enzymes can be formed rapidly. For example, when an organisms finds itself limited to an unusual carbon or nitrogen compound as sole source of carbon and energy, the synthesis of the enzymes required are normally repressed by a repressor, which prevents transcription of the DNA message to the messenger RNA. The inducer inactivates the repressor and thus makes transcription possible (Jacob and Monod, 1961).

When the microbial cell is faced with more than one utilizable substrate, it has to make a choice. If it would produce all the enzymes necessary for the utilization of all the substrates present, this would be less economic than producing enzymes for the utilization of one substrate after the other. The cell thus produces enzymes to utilize the best substrate present first and only after the exhaustion of this primary substrate are the enzymes formed for the next substrate. This phenomenon is called catabolite repression and referred to as diauxie.

Both of these mechanisms control the degradation of substrates at the cell membrane level and thus are also responsible for the uptake of the substrate. Since the substrate taken up by the cell is used by the organism to provide energy for growth and biosynthesis of molecular compounds such as enzymes, a proviso has to be made for possible overproduction of energy, which would lead to heat dissipation and thus overheating of the cell. This type of energy regulation is most prevalent

in aerobic metabolism and has been quantified by Atkinson (1969) with his energy charge equation

$$\text{Energy charge} = \frac{[\text{ATP}] + \frac{1}{2}[\text{ADP}]}{[\text{ATP}] + [\text{ADP}] + [\text{AMP}]}$$

This formula not only regulates the activities of the catabolic enzymes but also the biosynthetic enzymes that utilize ATP. The existence of such a control again indicates the coordination of control between catabolism or substrate utilization and growth or biosynthesis of enzymes and other growth-requiring organic compounds.

Similar to the controls of the substrate utilization, biosynthesis of the macromolecules from metabolic intermediates have also to be controlled. Apart from the energy regulation, biosynthesis has in addition feedback regulation. This type of regulation is in principle very similar to the one described for substrate utilization (catabolite regulation). The main difference is, of course, that the final metabolite of a biochemical sequence inhibits the action of an early enzyme of that sequence. It is therefore common to refer to feedback inhibition and feedback repression. Both mechanisms are required to adjust the rate of production of pathway endproducts to the rate of synthesis of macromolecules. Similar to the substrate utilization control in the presence of more than one substrate, special feedback regulatory systems exist in the case when more than one endproduct arises from a common metabolic sequence that branches at one or more points (Elander and Demain, 1981).

In addition to these metabolic regulations, microbial process development has to concern itself with one further selective mechanism often neglected or forgotten — permeability control. Metabolic regulation prevents over-synthesis of metabolites and macromolecules essential for the cell, whereas the permeability control or barrier by the cytoplasmic membrane makes certain of selectively bringing into the cell low molecular weight nutrients required and also of retaining concentrated solutions formed via metabolic events. It is therefore important to be familiar with the four main mechanisms of cell membrane transport (Wang *et al.*, 1979) and to know which one of them dominates in the particular microorganism considered for process development. Permeability can be changed or influenced by changes in the environment, whereby leakages can be instigated. The permeability barrier is also responsible for the problems in substrate utilization, if the available

substrate consists of polymers such as cellulose, starch or others. In these cases consideration has to be given to exoenzyme production by the micro-organism.

Whichever microbial process development is anticipated, any product formation — apart form biomass or single-cell protein — could therefore involve either an interference in the regulatory mechanisms, or mutation and thus interference in the genotype. Before this step is taken, however, one should have explored the phenotypic expression through environmental changes. Many primary products of microbial metabolism, in particular endproducts of catabolic events, can be obtained without changing the genotype but rather by improving the environmental and particularly the cultivation techniques. It is the latter which can in may instances overcome catabolite repression caused by these various endproducts of catabolism.

MODIFICATION OF THE GENETIC STRUCTURE OF THE ORGANISM TO INCREASE PRODUCT FORMATION

Intermediates of the degradative pathways or any compound connected with the biosynthesis of macromolecules, which also includes the secondary products, require the use of mutation techniques.

In order to understand fully the nature of mutagenesis it is necessary to have a general idea of the manner in which genetic information is encoded and deciphered into gene products (Jacobson, 1981). A mutation is any heritable change in the base pair sequence of an organism's DNA and is defined by any detectable, heritable change in an organism's phenotype. Changes within a coding region, for example, can alter the amino acid sequence of an enzyme and thereby effect its activity. This could lead to so-called auxotrophic mutants. With the removal of one enzyme from the biosynthetic sequence, the intermediate or substrate for this particular enzyme accumulates as a product. In order to ensure the continuation of growth and thus product formation, the would-be product of this enzyme or any of the subsequent intermediates must be fed into the fermentation vessel. The slight alteration of a promoter sequence could also increase the probability that RNA polymerase will bind to the promoter and thus enhance the rate of transcription. This is very important in the case of enzyme production. Mutations in operator regions or in a regulatory gene can prevent the binding of a repressor and thereby greatly increase

transcription. These so-called regulatory mutants are frequently used in amino acid, vitamin, etc., production, as they remove feedback regulation.

There exist several ways to bring about such changes or induce mutagenesis. Chemical mutagens can induce genetic changes by the direct induction of base mispairing, whereas physical agents damage DNA and thus can also cause gene mutations. Mutagens hit genes at random, which makes it impossible to cause a particular gene to mutate preferentially. To improve a strain by mutation one therefore has to rely on sensitive tests that make it possible to recognize and select the rare mutants which have the desired characteristics.

Auxotrophic and regulatory mutants together with parents strains are still the most commonly used microorganisms in industrial applications. Microbial process development is mainly concerned with selecting first the type of organism necessary to either develop a new or improve existing processes.

Strain improvement can also be achieved by hybridization, which is any crossing of genetically different individuals leading to an offspring with a genotype different from that of either parent (Esser and Stahl 1981). Hybridization is therefore a recombination of the genetic material. Here one has to distinguish between the sexual hybridization of eukaryotes and the parasexual hybridization of prokaryotes, imperfect eukaryotes and eukaryotic tissue cultures. Heterogenic incompatibility, however, restricts the use of this method and one should be aware that one cannot expect that fresh genetic information taken from nature by screening techniques will be suitable for the purpose of strain improvement by hybridization. In the case of parasexual processes, mechanisms known as conjugation, transduction, transformation and mitotic recombination have been used very frequently to bring the genetic material together. The barriers of homogenic and heterogenic incompatibility have been overcome to a large extent using the techniques of protoplast fusion and protoplast infection with DNA, which then lead to a whole new field of gene technology or genetic engineering.

Mutation alters a microorganism's genes, whereas reconbination rearranges genes or part of genes and brings together in an individual organism genetic information from two or more organisms.

The new and fast developing area of gene technology has its basis in the improvement of recombinant DNA techniques. The first process

was that of transferring plasmids. Plasmids are small circular molecules of extra-chromosomal DNA found in bacteria and yeasts, that are capable of autonomous replication within a cell and are inherited by daughter cells. These plasmids often carry genes that give certain bacteria specialized properties. They can be transferred from one bacterial strain to an unrelated strain, and sometimes to a different species, to introduce totally different new genetic properties to that bacterium. This ability to isolate plasmid DNA from a culture and induce another culture to take it up is the basis of most recombinant DNA manipulation.

The improvement of recombinant DNA techniques over the past decade has led to the use of plasmids as gene carriers and has opened the area for what is now known as genetical engineering. A gene or genes taken from an unrelated organism or an artificially synthesized gene can now be spliced into a plasmid and the plasmid introduced into a new microbial cell. The plasmid serves therefore as a vector for genes that have no counterpart in the recipient organism and could not be stably inherited in it through other recombinant techniques such as hybridization (Puhler and Heumann, 1981). The enormous development of gene technology or genetic engineering was made possible mainly because of the discovery of those enzymes which cause the above mentioned restrictions in gene modification, protecting the species against foreign genetic information. These so-called restriction endonucleases are now used

- to cleave a given DNA into characteristic fragments
- to split foreign DNA into a vector molecule such as the plasmid.

There exists a special system of nomenclature for the large number of individual restriction endonucleases. The second discovery was the ligating enzymes, which are capable of rejoining DNA fragments. This gene technology was soon extended from the prokaryotes to eukaryotic systems, and the use of plasmid vectors further to viruses as vectors, particularly in the case of animal and plant cells.

Recombinant DNA techniques can be applied in various ways for a number of different industrial purposes. The most widely known objective is the production by a microorganism of a protein it does not normally synthesize, such as an enzyme or a hormone. The idea is to transfer an individual gene coding for the desired product into a host microorganism and grow this microorganism in large volume to yield

the product. A different approach or objective of gene technology is the genetic improvement of an existing strain. Instead of introducing a brand-new genetic capability one can improve the efficiency of an existing strain by modifying its genetic information. Finally, this technology will make it possible to improve the precision of a more traditional approach by bringing about the mutation of specific sites in particular genes, overcoming the random nature of normal mutagenesis.

The generalized scheme in Figure 1 outlines the approach taken to date in obtaining cultures for microbial process development of a certain specified product. This does not mean, however, that large-scale processing can now commence. The modified strains using the various types of genetic modification have again to be submitted to the earlier described optimization studies. It is completely wrong to assume that the modified strain would behave in the established medium in exactly

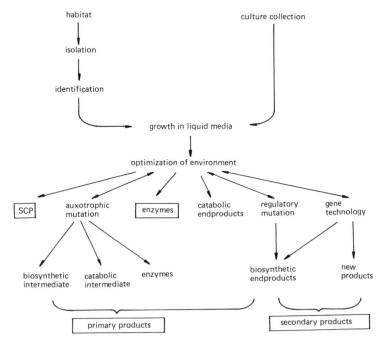

Fig. 1. General outline of microbial process development work for single-cell (biomass) or product formation processes.

the same way as the parent strain. The isolation and growth optimization may have been carried out in terms of the available substrate utilization, but the genetic manipulation in terms of product formation. Both aspects have now to be brought under one and the same optimization condition. Auxotrophic mutants may require the addition of growth factor compounds, regulatory mutants and genetically engineered microorganisms may also have some additional requirements. Whatever the chemical composition of the medium, it is imperative that all the components be thoroughly mixed, so that the microorganism or cell has ready access to the available nutrients and to the substrate.

CELL CULTIVATION SYSTEMS

With the selected microorganism and the growth medium finally developed, it is now necessary to look at the various types of cultivation techniques available, and select the technique which gives the best ratio between substrate utilization and product formation under the best economic conditions. Hereby one has to take into consideration the total sequence of steps in the industrial application of a microorganism as a biological catalyst as is outlined in Figure 2.

Since growth is normally considered as an increase in all cell material expressed in terms of cell mass or cell number, the growth process must therefore follow the overall kinetics of the integrated enzymatic activity, provided unlimited growth prevails indicating that all growth parameters (physical, chemical, biological) are optimized. The kinetics of microbial reactions (Pirt, 1975; Riviere, 1975; Doelle, 1975; Fiechter, 1981) in practical application strongly depends therefore on the extracellular parameters of the system. The selection of the entire cultivation equipment, which includes instrumentation for measurement, and central and peripheral parts, is tied to the cultivation method (Tannen and Nyiri, 1979; Fiechter, 1981).

In a *batch process* most or all of the constituents of the medium are combined with the biological catalyst at the start. In such a culture, microbial activity expresses the levels of mass transformation and kinetics is governed by a large number of intracellular and extracellular parameters. Since these parameters act on the system in a strongly time-dependent manner, the expression of life is a dynamic process. For improved efficiency in cultivation method and also in industrial application, methods for the proper identification and measurement of the

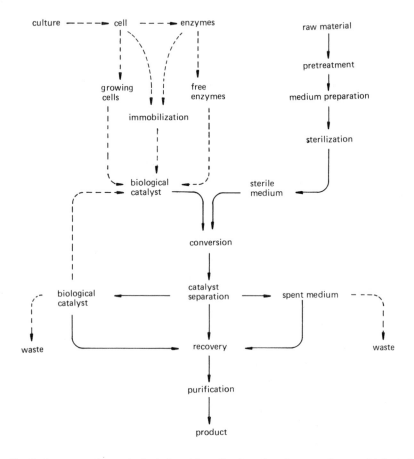

Fig. 2. Sequence of steps in the industrial application of a microorganism as a biological catalyst varies from one process to another, but follows the general outline scheme (adapted from Gaden 1981).

relevant parameters are necessary, whereby the primary problem of proper control is the process identification. In such a batch culture, the inoculated culture is grown, the fermenter is emptied after completion and the required product extracted. Such a process is characterized by

$$\text{rate of growth } (\mu) = \frac{1}{X}\,\frac{\mathrm{d}X}{\mathrm{d}t}$$

$$\text{rate of production} = \frac{1}{X}\,\frac{\mathrm{d}P}{\mathrm{d}t}$$

$$\text{rate of substrate utilization} = \frac{1}{X}\left(-\frac{\mathrm{d}S}{\mathrm{d}t}\right)$$

During the exponential phase, the growth reaches its maximum value μ_{max}, if all parameters are optimal. This maximal value can only be reached therefore during unlimited growth. If a single substrate exerts a limiting effect, the growth rate will be less than μ_{max}. It is therefore possible to determine the concentration of the limiting substrate through the relationship

$$\mu = \mu_{max}\,\frac{S}{Ks + S}$$

whereby Ks is that value of the substrate concentration at which the growth rate is half the maximum growth rate μ_{max} (Doelle, 1975; Pirt, 1975; Wang *et al.*, 1979). Since the substrate concentration is directly correlated to the number of cells formed, the yield factor together with Ks and μ_{max} allows the calculation of substrate required to obtain maximal biomass

$$Y_S = \frac{g\ \text{biomass formed}}{g\ \text{substrate used}}\,\frac{\mathrm{d}X}{\mathrm{d}S}$$

This equation can be extended, of course, to specific atoms such as carbon, nitrogen, phosphorus, sulphur and others, depending upon the process under development.

The second cultivation technique is the *continuous cultivation* of micro-organisms, which has the potential of a higher volume of production for an installation of a given size and may have considerable advantages over sequential batch fermentations. The simplest approach is to modify a batch reactor so that fresh nutrient and substrate can continually be added and the products of the reaction can continually be removed. Depending upon the type of control, one distinguishes between turbidostat and chemostat. In the first case, the turbidity yields a measure of the rate at which cells leave the tank and adjusts the inflow of nutrients accordingly. The second method of controlling a continuous culture is much simpler and can be applied in cases where the product of the reaction does not consist of cells. In this case the reactor is

controlled by monitoring the input rather than the output stream. The concentration of a critical nutrient is fixed at a level such that the other nutrients are abundant. The level of this nutrient then limits the extent to which the microorganisms proliferate. If one keeps the flow rates, the working volume and all other parameters constant, a steady state of population characteristics will be reached. In taking flow kinetics into our earlier mentioned growth kinetics,

$$\frac{dX}{dt} = \mu X - DX$$

whereby the dilution rate D relates the flow rate F to the actual volume V of the reactor:

$$D = \frac{F}{V}$$

At the steady state, the increase in biomass per unit time is zero and hence $D = \mu$. The steady state conditions can be achieved for various values of D and μ varying from almost zero to a maximum, which is characterized by the genotype and μ_{max} value of the particular micro-organism. The cultivation technique has a self-regulatory capacity between the two extremes, provided the above mentioned basic conditions for continuous cultivation are obeyed. Similar to the batch fermentation, a correlation exists between X, S, μ and D:

$$X = Y\left(So - \frac{KsD}{\mu_{max} - D} \right)$$

Continuous cultivations can be carried out in single-stage and multi-stage chemostat systems (Fiechter, 1981). The major disadvantage is, however, that the use of the continuous cultivation requires either a process for biomass production or for products which are growth-associated.

Fed-batch cultivation is a particular form of continuous cultivation consisting of a continuous or sequential addition of medium to an initial batch, without any withdrawal of reaction mixture. This means, that the actual volume V in the reactor is not constant and varies according to the volume F added:

$$F = \frac{dV}{dt} \quad \text{(Fiechter, 1981)}$$

A fed-batch culture system therefore never attains a steady state, but rather a quasi-steady state (Pirt, 1975). If the Monod-type growth characteristics are reached,

$$\frac{dX}{dt} = (\mu - D)X$$

and the rate of increase of total biomass,

$$\frac{dX}{dt} = FYS_0;$$

whereby $X_{max} = X_o + FYS_o t$.

Comparing these equations with those of the continuous cultivation system, it can easily be deduced that the main difference lies in the changing value for the actual volume V. These fed-batch culture systems can be of great importance for overcoming substrate inhibition.

A rather new approach is the cultivation of microorganisms by *dialysis*. In this technique, the apparatus consists essentially of two compartments separated by a membrane. One compartment maintains the microorganism in suspension in a liquid medium, whereas the other, generally much larger vessel, contains fresh medium. Nutrients can now pass across the membrane to the reactor vessel and metabolites are able to pass to the nutrient vessel. The regulation of the system depends on the properties of the membrane (Riviere, 1975) and the ratio of the volume of the two compartments. Dialysis cultures are claimed to have two special characteristics. Firstly, one observes an extension of the exponential phase during the batch fermentation, which may be due to the continued supply of nutrients and the removal of toxic or inhibitory metabolites. Secondly, a considerable increase of biomass concentration has been observed. The dialysis culture technique would have the great advantage to overcome catabolic intermediate or endproduct inhibition. It can also facilitate the recovery of macromolecules.

All of these very briefly mentioned types of cultivation techniques have been developed to overcome certain problems in biomass or product formation, not only using microorganisms, but also in the case of plant and animal cell cultivation. Many variations exist of each of these systems and there is no doubt in my mind that many more variations will be designed in the future. It is one of the most difficult problems in microbial process development to find or design a cultivation technique which results in the eventual goal, the most economic

conversion of substrate to product. An ideal cultivation technique should also minimize the recovery of the product (Belter, 1979) and minimize wastes. A very dangerous and also costly waste or pollutant in the outflowing medium is the catalyst itself. Firstly, every microorganism has to be regarded as a potential health hazard and thus should be destroyed before the effluent is released. This may be a very costly undertaking in industrial plants. Secondly, the value of a biological catalyst is at least equal to the value of the nutrients consumed in growing the cells, which means that its waste also can be very costly (Gaden, 1981).

There are at the moment two ways to limit the amount of catalyst that is lost. One way is to recycle the catalyst from the overflow back into the reactor. Such recycling systems have been tried, but turn out to be a difficult and costly system particularly in the case of bacteria, since the cells are often damaged and the recycling system is open to contamination with foreign microorganisms. The second alternative is, of course, to keep the catalyst in the reactor. The most common technique under investigation and also employed is the packed bed system, which uses a solid support on which the cells are encouraged to grow.

Methods for such an immobilization of microbial cells (Chibata *et al.*, 1979) can be classified into three categories analogous to the immobilization of enzymes (Wang *et al.*, 1979). The simplest method is the carrier-binding one, which is based on the direct binding of microbial cells to water-insoluble carriers. The binding is ionically of nature using carriers containing ion-exchange residues. On autolysis, cells leak out from the carrier, which results in a loss of catalyst, but neverthless a cleaner effluent. A cross-linking with bi- or multi-functional reagents such as glutaraldehyde is another method of microbial cell immobilization, but little success has been obtained up-to-date. A much more popular immobilization method is the one which directly entraps microbial cells into polymer matrices. Collagen, gelatin, agar, alginate, carrageenan, polyacrylamide, polystyrene and others have been used. When these polymers are permeated with water, they form a meshwork of fibres with ovids, where cells or enzymes can be entrapped. A limitation of this technique is, however, that nutrients and product must diffuse through the solid matrix, which can reduce the rate of reaction significantly. On the other hand, the catalyst is held firmly in place without damage. Microencapsulation is a further technique for immobilization, whereby cells or enzymes are enclosed in a spherical polymer

and semipermeable membrane. The resulting capsules range in diameter from 5 to 300 micrometers and look like enlarged cells. This method allows small molecule nutrients and products to pass through the membrane freely, but holds back the larger molecules such as cells and enzymes. There is no doubt that this promising new development requires a tremendous amount of research to make these techniques much more efficient.

One of the final considerations to be given in microbial process development is the available substrate. Most of the cheap raw materials are not monomeric but rather polymeric in nature, which means that either physical or chemical pretreatment is required, or the cell must induce extracellular enzymes to break down these polymers to their individual monomers. These problems lead to another new field of enzyme production, enzyme engineering and enzyme immobilization (Wang et al., 1979). It is in the hands of the biotechnologists to decide, whether it is more economic to carry out

(1) physical or chemical pretreatment;
(2) mixed population systems, where one organism breaks down the polymer and the second converts the monomer to the desired product;
(3) microbial gene manipulation to implant into the microorganism the additional gene(s) necessary to break down the polymer;
(4) enzyme hydrolysis, using pure enzymes from an additional enzyme production plant (Aunstrup et al., 1979);
(5) enzyme hydrolysis, using immobilized enzyme systems for the pretreatment.

At the present stage of development, the choice is very limited, as most of the enzymes involved are either still unknown, unstable or very expensive to produce.

The development of microbial processes as has been described is usually carried out at the bench scale, where basic screening procedures are carried out and at pilot plant scale to ascertain optimal operating conditions. It is now time to carry the results of the laboratory and pilot plant scale investigations to the chemical engineering section for translation to the plant scale. Computers (Armiger and Humphrey, 1979), microcomputers (Hampel, 1979), statistical models (Ramkrishna, 1979), mass and energy balances (Nagai, 1979) are the basis for plant

design, process design and the evaluation of the economics of large scale fermentation processes (Bartholomew and Reisman, 1979).

FUTURE OF MICROBIAL PROCESS DEVELOPMENT

A tremedous amount of literature is available on microbial processes, with the most economic ones falling probably into antibiotic and amino acid production (Hirose and Okada, 1979; Perlman, 1979). Most processes, however, rely mainly on old traditionally used microorganisms and technology with very little of the necessary comprehensive reinvestigation of the catalyst being involved. The reason for their enormous publicity was caused by the energy crisis and the hopes for quick answers, whereby reference is made particularly to the ethanol production processes, which all are still very uneconomical and pollution prone (Bungey, 1981; Maiorella *et al.*, 1981; Kosaric *et al.*, 1981; Rogers *et al.*, 1982). It is encouraging to see the upsurge in research in biomass conversion programmes all over the world (DaSilva, 1980; Carioca *et al.*, 1981; Linko, 1981; Sahm, 1981; Potgieter, 1981; Haltmeier, 1981; Eriksson, 1981; Hammond, 1982), But unfortunately all these programmes still lack comprehensive microbial process development and rely on cultures from culture collections or information from basic academic research for plant scale development. Present biotechnology development lean heavily on the chemical engineering rather than the microbiological aspect. The reasons for this are undoubtedly due to the enormous experiences gained in biochemical engineering in the 1970s in the case of animal feed or single-cell protein production. On the other hand, there is no doubt that the most intensive microbial process development probably occurs at present in the field of agricultural microbiology (Brill, 1981), that is, nitrogen fixation.

Most scientists, biotechnologists and particularly government agencies seem not to be aware as to the potential and urgent need for basic and applied microbial process development. How can one design an efficient plant without having an efficient catalyst for the transformations required? One has to explore the potential of the microbe first before a decision can be made whether or not the particular microbe is suitable for process development. This, of course, involves long-term and unfortunately not short-term planning and research.

The future of biotechnological reasearch and development moves basically into two directions: firstly the well-known single-product

formation (SCP, amino acids, antibiotics, etc), and secondly into the new era of socio-economics, that is, integrated technology (DaSilva *et al.*, 1978; 1980; Doelle and DaSilva, 1980; Doelle, 1982b) using renewable or bioresources as substrates (King *et al.*, 1980). Biogas or methane (Scharer and Moo-Young 1979; Hobson *et al.*, 1981) is probably one of the most popular products at the present time, but others are described elsewhere (Olguin, Chiao, this volume). In this area of complete exploitation of the substrate to products and clean effluent lies the greatest challenge for microbial process developers.

REFERENCES

Aharonowitz, Y. and Cohen, G. (1981). 'The Microbiological Production of Pharmaceutical.' *Sci. American* **245**, 106−109.

Armiger, W. B. and Humphrey, A. E. (1979). 'Computer Application in Fermentation Technology.' In *Microbial Technology* (H. J. Peppler and D. Perlman, eds.). 2nd ed., Vol. 2, pp. 375−401. Academic Press, New York.

Atkinson, D. E. (1969). 'Regulation of Enzyme Function.' *Ann. Rev. Microbiol* **23**, 47−68.

Aunstrup, K., Andresen, O., Falch, E. A. and Nielsen, T. K. (1979). 'Production of Microbial Enzymes.' In *Microbial Technology* (H. J. Peppler and D. Perlman, eds.), 2nd ed., Vol. 1, pp. 282−310. Academic Press, New York.

Bartholomew, W. H. and Reisman, H. B. (1979). 'Economics of Fermentation Processes.' In *Microbial Technology* (H. J. Peppler and D. Perlman, eds.). 2nd ed., Vol. 2, pp. 463−490. Academic Press, New York.

Belter, P. A. (1979). 'General Procedures for Isolation of Fermentation Products.' In *Microbial Technology* (H. J. Peppler and D. Perlman, eds.) 2nd ed., Vol. 2, pp. 403−432, Academic Press, New York.

Brill. W. J. (1981). 'Agricultural Microbiology.' *Sci. American* **245**, 146-156.

Bull, A. T., Ellwood, D. C. and Ratledge, C. (1979). 'The Changing Scene in Microbial Technology.' *Symp. Gen. Microbiol.* **29**, 1−28.

Bungay, H. R. (1981). 'Biochemical Engineering for Fuel Production in the United States.' *Adv. Biochem. Eng.* **20**, 1−14.

Campoz-Lopez, E. (1980). *Renewable Resources: A Systematic Approach.* Academic Press, New York.

Carioca, J. O. B., Arora, H. L. and Khan, A. S. (1981). 'Biomass Conversion Program in Brazil.' *Adv. Biochem. Eng.* **20**, 153−162.

Chatel, B. (1980). 'Bioresources for Development.' In *Bioresources for Development* (A. King, H. Cleveland and G. Streatfield, eds.), p. 51. Pergamon Press, New York.

Chibata, I., Toas, T and Sato, T. (1979). 'Use of Immobilized Cell Systems to Prepare Fine Chemicals.' In *Microbial Technology* (H. J. Peppler and D. Perlman, eds.), 2nd ed., Vol. 2, pp. 434−462.

Cooney, C. L. (1981), 'Growth of Microorganisms.' In *Handbook of Biotechnology* (H. Rehm and G. Reed, eds.) 1, pp. 73−113. Verlag Chemie, Weinheim.

DaSilva, E. J. (1981). 'The Renaissance of Biotechnology: Man, Microbe, Biomass and Industry.' *Acta Biotechnol.* **1**, 207—246.

DaSilva, E. J., Olembo, R. and Burgers, A. (1978). 'Integrated Microbial Technology for Developing Countries: Springboard for Economic Progress.' *Impact of Science on Society* **28**, 159—181.

DaSilva, E. J., Shearer, W. and Chatel, B. (1980). 'Renewable Bio-Solar and Microbial system in "Eco-Rural" Development.' *Impact of Science on Society* **30**, 225—233.

Demain, A. L. and Solomon, N. A. (1981). 'Industrial Microbiology.' *Sci. American* **245**, 42—51.

Dietz, A. (1981). 'Pure Culture Methods for Industrial Microorganism.' In *Handbook of Biotechnology* (H. J. Rehm and G. Reed, eds.) 1, pp. 411—434, Verlag Chemie, Weinheim.

Doelle, H. W. (1975). *Bacterial Metabolism.* 2nd ed., Academic Press, New York.

Doelle, H. W. (1981). 'Basic Metabolic Processes.' In *Handbook of Biotechnology* (H. J. Rehm and G. Reed, eds.) 1, pp. 113—210. Verlag Chemie, Weinheim.

Doelle, H. W. (1982). Editorial in *IOBB Newsletter* 'Biotechnologia', April issue.

Doelle, H. W. (1982b). 'Appropriate Biotechnology for Developing Countries.' *Conservation and Recycling* **5**, 75—77.

Doelle, H. W. and DaSilva, E. J. (1980). 'Microbial Technology and Its Potential for Developing Countries.' *Process Biochem* **15**(3), 2—6.

Elander, R. P. and Demain, A. L. (1981). 'Genetics of Microorganisms in Relation to Industrial Requirements.' In *Handbook of Biotechnology* (H. J. Rehm and G. Reed, eds.) 1, pp. 235—278. Verlag Chemie, Weinheim.

Eriksson. K. E. (1981). 'Swedish Development in Biotechnology Based on Ligno-Cellulosic Materials.' *Adv. Biochem. Eng.* **20**, 193—203.

Esser, K. and Stahl, U. (1981). 'Hybridization.' In *Handbook of Biotechnology* (H. J. Rehm and G. Reed, eds.) 1, pp. 305—330. Verlag Chemie, Weinheim.

Eveleigh, D. E. (1981). 'The Microbial Production of Industrial Chemicals.' *Sci. American* **245**, 120—133.

Fiechter, A. (1981). 'Batch and Continuous Culture of Microbial, Plant and Animal Cells.' In *Handbook of Biotechnology* (H. J. Rehm and G. Reed, eds.), 1, 453—505. Verlag Chemie, Weinheim.

Gaden, E. L. (1981). 'Production Methods in Industrial Microbiology.' *Sci American* **245**, 134—144.

Haltmeier, T. (1981). 'Biomass Utilization in Switzerland.' *Adv. Biochem. Eng.* **20**, 189—192.

Hammond, K. W. (1982). *Biotechnology: Preparing for a Bio-society.* Conference Report Club of Guelph, Univ. Guelph, Ontario, Canada.

Hampel, W. (1979). 'Application of Microcomputers in the Study of Microbial Processes.' *Adv. Biochem. Eng.* **13**, 1—34.

Hirose, Y. and Okada, H. (1979). 'Microbial Production of Amino Acids.' In *Microbial Technology* (H. J. Peppler and D. Perlman, eds.), 2nd ed., 1, 211—240. Academic Press, New York.

Hobson, P. N., Bonsfield, S. and Summers, R. (1981). *Methane Production from Agricultural and Domestic Wastes.* Appl. Science Publ., London.

Hopwood, D. A. (1981). 'The Genetic Programming of Industrial Microorganisms.' *Sci. American* **245**, 66—93.

62

H. W. Doelle

Jacob, F. and Monod, J. (1961). 'Genetic Regulatory Mechanisms in the Synthesis of Proteins.' *J. Mol. Biol.* **3**, 318–356.

Jacobson, G. K. (1981). 'Mutations.' In *Handbook of Biotechnology* (H. J. Rehm and G. Reed, eds.), 1, pp. 279–304. Verlag Chemie, Weinheim.

King, A., Cleveland, H. and Streatfield, G. (1980). *Bioresources for Development: The Renewable Way of Life.* Pergamon Press, New York.

Kockova-Kratockvilova, A. (1981). 'Characteristics of Industrial Microorganisms.' In *Handbook of Biotechnology* (H. J. Rehm and G. Reed, eds.) **1**: 5–71, Verlag Chemie, Weinheim.

Kosaric, N., Durnjak, Z. and Stewart, J. J. (1981). 'Fuel Ethanol from Biomass: Production, Economics and Energy.' *Adv. Biochem. Eng.*, 119–151.

Linko, M. (1981). 'Biomass Conversion Program in Finland.' *Adv. Biochem. Eng.* **20**, 163–172.

McGowan, V. F. and Skerman, V. B. D. (1982). *World Directory of Collections of Cultures of Microorganisms.* 2nd ed., Brisbane.

Maiorella, B., Wilke, C. R. and Blanch, H. W. (1981). 'Alcohol Production and Recovery.' *Adv. Biochem. Eng.* **20**, 43–91.

Nagai, S. (1979). 'Mass and Energy Balance for Microbial Growth Kinetics.' *Adv. Biochem. Eng.* **11**, 49–84.

Nakayama, K. (1981). 'Sources of Industrial Microorganisms.' In *Handbook of Biotechnology* H. J. Rehm and G. Reed, eds.), 1, pp. 355–410. Verlag Chemie, Weinheim.

Perlman, D. (1979). 'Microbial Production of Antibiotics.' In *Microbial Technology* (H. J. Peppler and D. Perlman, eds.) 2nd ed., Vol. 1, pp. 241–280. Academic Press, New York.

Pirt, J. (1975). *Principles of Microbe and Cell Cultivation.* Blackwell Scientific Publ., London.

Porter, J. R. (1980). 'Microorgamisms as Bioresource Potentials for Development.' In *Bioresources for Development* (A. King, H. Cleveland and G. Streatfield, eds.), pp. 12–32. Pergamon Press, New York.

Potgieter, H. J. (1981). 'Biomass Conversion in South Africa.' *Adv. Biochem Eng.* **20**, 181–188.

Puhler, A. and Heumann, W. (1981). 'Genetic Engineering.' In *Handbook of Biotechnology* (H. J. Rehm and G. Reed, eds.), 1, pp. 331–354. Verlag Chemie, Weinheim.

Ramkrishna, D. (1979). 'Statistical Models of Cell Populations.' *Adv. Biochem. Eng.* **11**, 1–48.

Rogers, P. L. (1978). 'Microbial Process Technology for Developing Countries.' In *GIAM V* (O. K. Stanton and E. DaSilva, eds.), p. 228 UNEP/Unesco/ICRO panel, Kuala Lumpur.

Rogers, P. L., Lee, K. J., Skotnicki, M. L. and Tribe, D. E. (1982). 'Ethanol Production by Zymomonas mobilis.' *Adv. Biochem. Eng.* **23**, 37–84.

Rose, A. H. (1981). 'The microbiological production of food and drink.' *Sci. American* **245**, 94–105.

Rivière, J. la (1975)*Industrial Applications of Microbiology.* Surrey Univ. Press, London.

Sham, H. (1981). 'Biomass Conversion Program of West Germany.' *Adv. Biochem. Eng.* **20**, 173–180.

Scharer, J. M. and Moo-Young, M. (1979). Methane Generation by Anaerobic Digestion of Cellulose-Containing Wastes. *Adv. Biochem. Eng.* **11**, 85−101.

Sly, L. and Atthasampunna, P. (1984). *Culture Collection Technologies in Conserving the Planet's Microbial Heritage.* Trends in Appl. Microbiol. Biotechnol.

Tannen, P. L. and Nyiri, L. K. (1979). 'Instrumentation of Fermentation Systems.' In *Microbial Technology* (H. J. Peppler and D. Perlman, eds.) 2nd ed., Vol. 2, pp. 331−376. Academic Press, New York.

Veldkamp, H. (1970). In *Methods in Microbiology* (J. R. Norris and R. W. Ribbons, eds.) Vol. 3, pp. 305−361. Academic Press, New York.

Wang, D. I. C., Cooney, C. L., Demain, A. L., Dunnill, P., Humphrey, A. E. and Lilly, M. D. (1979). *Fermentation and Enzyme Technology.* J. Wiley & Sons, New York.

Biotechnology Group,
Department of Microbiology,
University of Queensland,
St. Lucia, Qld 4067,
Australia.

M. Ringpfeil and B. Heinritz

Single-Cell Protein Technology

INTRODUCTION

The industrial production of single-cell protein (SCP) is the first attempt in the history of mankind to produce proteinaceous food and feedstuffs without the aid of agricultural means, e.g. photosynthesis and extended areas. The necessary energy source is made available by reduced carbon compounds of fossil or renewable origin.

There is a competition between the industrial and the agricultural production of food and feedstuffs. The advantages of agricultural production lie in the use of cheap energy (sun) and carbon (CO_2) sources. The advantages of microbial production lie in the industrial organization of the production, its independence from climate and temperature, and its limited demand on space and area. Microbial production will compete successfully if the following inequality is fulfilled:

(1)	Cost of food or feedstuffs of animal or vegetable origin	$>$	Cost of carbon-containing raw material for microbial production	$+$	Cost of production of microbial food or feedstuffs

Another possibility of producing SCP by means of phototrophic microorganisms gives rise to a different type of competition with conventional agricultural production, in which the same energy and carbon sources are used, but which is based on aqueous ponds of natural or artificial origin instead of soil. The description of SCP technology given below is limited to methods using reduced carbon compounds.

THE PROPERTIES OF SCP IN APPLICATION

SCP used in the form of dried cells or protein concentrates or isolates derived from whole cells serves as an additive to food and feedstuffs due to its high protein content. At present it is used mainly for providing

64

protein to animal feedstuffs, though it is also intended to be used in the future for humans.

There are some further SCP applications of minor importance which make use of its abilities to improve such properties as water- and fat-binding, emulsion stability, palatability, whippeability, and geleability of certain foodstuffs (Shay, personal communication, 1978).

The amino acid composition of SCP does not differ significantly from those of proteins of vegetable or animal origin (Table I). However, certain differences in the quantitative composition may be observed between them, as well as between SCP derived from different micro-organisms.

For optimum nutrition, the amount of SCP added to the food or feedstuff is calculated so that an optimum amino acid composition of the whole mixture may be obtained (Pokrowskij, 1972; Pokrowskij *et al.*, 1978; Taylor and Senior, 1978).

The nutritional value of food and feedstuffs is determined by their biological value and their digestibility, the biological availability of their amino acids, the net protein utilization (NPU) and the protein efficiency ratio (PER). The biological value of a protein corresponds to the quantity of retained nitrogen divided by the quantity of nitrogen uptake multiplied by 100. The digestibility of a protein corresponds to the quantity of absorbed nitrogen divided by the total nitrogen uptake multiplied by 100.

The biological availability of amino acids may be measured by special chemical, microbiological, or enzymatic methods. The net protein utilization is the product of the biological value, and the digestibility. The protein efficiency ratio is expressed by the increase of body weight divided by the weight of proteins taken up (Young and Scrimshaw, 1975; Blanc, 1978).

The application of SCP for food and feed is inevitably connected with considerations concerning the health of man. Consequently, tremendous efforts have been made in the past three decades to find out the effects of SCP on higher organisms such as target animals and even man himself. The whole work has been assessed by the Protein Advisory Group (PAG) of the UN system (1976) and directed by guidelines issued by the PAG as well as the Commission on Fermentation of the IUPAC (1974–1978). In conclusion it could be pointed out that SCP from fossil and renewable resources is absolutely safe for man so long as the necessary tests carried out are in agreement with

TABLE I

Amino acid content of some proteins (in grams per 16 grams of nitrogen)

| Amino acid | FAO standard[a] | Fish meal[b] | Extracted soya bean meal[b] | Single cell protein | | | | | | Microfungi |
| | | | | Yeasts | | Bacteria | | | | |
				Gas oil[c]	Fermosin[e]	Pruteen[d]	Methanol[e]	Methane[e]	Methane[d]	Pekilo[f]
Isoleucine	4.2	4.6	5.4	4.0	6.0	4.7	3.8	4.3	4.8	4.3
Leucine	4.8	7.3	7.7	6.8	8.6	7.2	8.0	8.1	7.6	6.9
Phenylalanine	2.8	4.0	5.1	3.8	5.2	3.6	3.9	4.6	4.0	3.7
Threonine	2.8	4.2	4.0	4.7	5.8	4.7	4.5	4.7	4.5	4.6
Valine	4.2	5.2	5.0	4.7	6.3	5.7	5.3	6.5	6.2	5.1
Cysteine	2.0	1.0	1.4	0.9	0.8	0.7	0.6	0.6	0.5	1.1
Methionine	2.2	2.6	1.4	1.6	1.6	2.4	2.1	2.7	2.2	1.5
Lysine	4.2	7.0	1.5	4.5	7.9	6.2	5.1	5.7	5.7	6.4

According to:
a Norris (1968).
b Shaklady (1975).
c Smith and Palmer (1976).
d Hamer et al. (1975).
e Faulhaber (unpublished results).
f Romantschuk (1975)

the IUPAC and PAG guidelines. In order to assure such safety, the PAG guidelines have been revised in 1982 under auspices of the UN University (PAG/UNU, 1982) and will be reissued shortly (Scrimshaw, personal communication).

GENERAL PROCEDURE OF SCP PRODUCTION

The basic steps in the general scheme of SCP production consists of the mixing of raw materials, fermentation, mechanical phase and thermal phase separation. Some variations may occur depending upon the use of different raw materials. Polymeric raw materials such as cellulose or starch necessitate a pretreatment step. Purified normal paraffins require a post-fermentative ripening step in order to reduce the amount of residual hydrocarbons to almost zero. Gas oil necessitates the mechanical separation of excess gas oil and the extraction of the remaining hydrocarbons by solvent treatment, which simultaneously makes it possible to produce valuable by-products. The use of natural gas gives an opportunity to use excess gases for heating purposes. Depending upon the type of raw material, the input auxiliary substances as well as the purification of effluents becomes necessary or advantageous (Fig. 1; Ringpfeil, 1978).

Additional treatment of the resulting SCP can be performed in order to produce more highly purified proteinaceous substances such as isoconcentrates (extracted cell protein precipitated on cell residues), concentrates (cells partly freed from non-proteinaceous substances) or isolates (extracted and purified cell protein) (Fig. 1; Sonnenkalb, unpublished results).

A deviation from this general scheme of SCP production has been developed in recent years by introducing the semisolid fermentation procedure. In general, polymeric solid materials such as plant materials are mixed with water and the necessary nutrients, and exposed to mycelium organisms. The resulting protein-enriched mixture is dried or used directly as feed (Senez, 1983). This procedure is very similar to certain fermentations applied in the production of indigenous foods (Steinkraus, 1980; Waslien and Steinkraus, 1980).

SCP production is carried out in large plants with an output of several thousand to several hundred thousand tonnes of dried cells per year. The largest plants are located in the U.S.S.R., which altogether produces more than one million tonnes of SCP per year (Ritshkov, 1982).

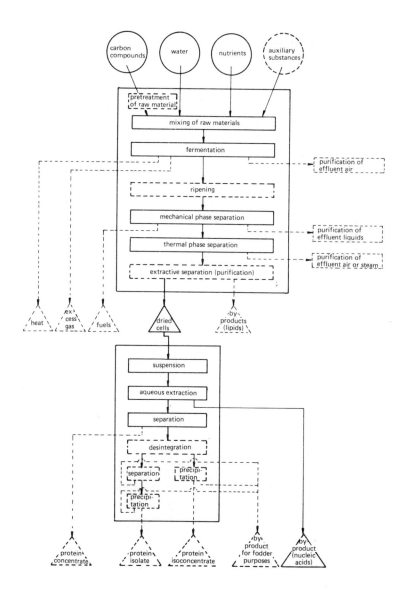

Fig. 1. General procedure of SCP production.

SCIENTIFIC BASIS OF PRODUCING AND RECOVERING SCP

The overall reaction equation of producing cell mass is the same for all types of substrates or microorganisms:

$$(2) \quad \left\{ \begin{array}{l} a_0 CH_{1.7}O_{0.5}N_{0.17}P_{0.01} \ldots + \\ \\ bCH_nO_m + cO_2 + dNH_4^+ + ePO_4^{3-} + \ldots \end{array} \right\}$$

$$\left\{ \begin{array}{l} (a - a_0)CH_{1.7}O_{0.5}N_{0.17}P_{0.01} \ldots + \\ \\ fCO_2 + gH_2O + \ldots \end{array} \right\} Q < O$$

A certain type of microorganism is growing quantitatively on a complex medium in which the carbon, oxygen, nitrogen and phosphorus sources are predominant. The amount of carbon source is of the same order as that of the resulting cell mass because approximately half of it is used by the growing microorganisms for providing them with biologically utilizable energy. Therefore, CO_2 and H_2O also predominate quantitatively on the product site.

Biologically utilizable energy is produced in the most efficient way by aerobic metabolism of carbon sources via oxidative phosphorylation (Mitchell, 1961; Lehninger, 1970; Skulatshew, 1969). Therefore, SCP production is preferentially carried out under aerobic conditions, although significant production of heat is unavoidable.

The overall growth reaction for microbial populations as well as for single cells cultivated on all microbially utilizable carbon and energy sources is of the autocatalytic type:

$$(3) \quad \frac{\dot{x}}{x} = \mu$$

This equation shows that the specific growth rate constant (μ) equals the quotient of the velocity of cell growth ($\dot{x}$) and the cell concentration (x). A solution of the equation represents the exponential growth law:

$$(4) \quad x = x_0 \cdot e^{\mu t}$$

$$(5) \quad \mu = \frac{\ln 2}{T}$$

A given amount of cells (x_0) doubles repeatedly at constant time intervals (T).

This exponential growth law behaviour of the microbial population was used empirically for continuous cultivation in an ideally mixed vessel well before this theoretical consideration was established (Rieche, 1954). The kinetic and hydrodynamic theories of cell growth (Malek, 1958; 1964) and the flow characteristics in reaction vessels (Jones, 1951) combined formed later the theory of continuous cultivation and confirmed the empirically elaborated continuous cultivation method for SCP production. This theory is still valid and widely used in the laboratory and in industrial practice:

(6) $\dfrac{dx}{dt} = Dx_0 + \mu x - Dx$

(6a) $Dx_0 = 0$

(6b) $\dfrac{dx}{dt} = 0$

(6c) $\mu = D$

The mass balance for an ideally mixed vessel, in which cell growth takes place, shows the equality of the dilution rate and of the specific growth rate (Equation (6c)) under steady-state conditions (Equation (6b)), assuming that no addition of microorganisms occurs via the media inflow (Equation (6a)). This theory has been applied most frequently in the chemostat system of continuous cultivation. Here a reactant limits the autocatalytic growth according to the Monod equation (Monod, 1949, 1950):

(7) $\mu = \mu_{max} \dfrac{s}{k_s + s}$

This equation is derived from the Michaelis-Menten equation which is valid for enzyme reactions and applied to the complex enzymatic reaction system of a microbial cell assuming that a single enzymatic reaction controls the overall reaction rate of growth.

The application of this equation is the 'weak point' of the theoretical background of continuous cultivation, although it is commonly used in practice today. Moreover, the Monod equation is mathematically equal to a simple approximation method. Therefore, the experimentally observed behaviour of the growth curve could be interpreted sufficiently by this method despite the real cause of the limitation. Although the value of k_s is always real for a real growth curve, it is often not known

which reactant is responsible for that value. Owing to this situation, the approximation of μ to μ_{max} prime, in order to find the most effective way of continuous cultivation, is still a time-consuming empirical effort, with misinterpretations of these empirical observations undoubtedly possible.

In most fermentations oxygen plays an extraordinary role due to its low solubility in water. Oxygen cannot be stored in the inflowing medium for the reaction. It must be supplied to the reaction mixture continuously. Therefore, in most of the practical cases today the oxygen transfer rate from the gas to the liquid limits the absolute value of the productivity of the fermentation.

For a long time SCP production was carried out at low productivities (Ringpfeil, 1981a). The systematic consideration of this behaviour gave rise to the development of high-performance oxygen transfer devices for fermentation (Liepe *et al.*, 1978), and the encouraging results obtaining high productivities (Ringpfeil, 1981a) in these devices led to reflections about maximum productivity (Ringpfeil, 1980). These considerations revealed the present uncertainty about the exact value of the maximum specific growth rate constant (μ_{max}). It is by no means comparable with the material constants known in chemistry. Therefore, the calculation of maximum productivities contains still many uncertainties.

The quotients of the coefficient of the resulting cell mass and the coefficients of every reactant (Equation (2)) are the yield values of cell mass referred to the specific reactant:

$$(8) \qquad \frac{a - a_0}{b} = Y_{x/s}; \quad \frac{a - a_0}{c} = Y_{x/c}; \quad \frac{a - a_0}{d} = Y_{x/NH_4^+};$$

$$\frac{a - a_0}{e} = Y_{x/PO_4^{3-}}; \; \ldots$$

$$(9) \qquad Y_{x/s} \sim Y_{x/c} \sim Y_{x/NH_4^+} \sim Y_{x/PO_4^{3-}} \cdots \sim \frac{1}{Y_{CO_2/x}} \cdots \sim \frac{1}{Y_{Q/x}}$$

During the course of a fermentation proportional changes of all yield values may occur.

In accordance with the economic aim of SCP production, maximum yield values are wanted for cell mass and, consequently, minimum yield values for CO_2, H_2O, and other reaction products including heat. The use of more expensive raw materials such as methanol, ethanol, sucrose and purified long-chain paraffins for SCP production gave rise

to further theoretical considerations in regard to maximum yield values. Thermodynamics was combined with specialized knowledge and certain assumptions on specific pathways of energy metabolism, which lead to indications for more exact maximum values (Payne, 1970; Stouthamer, 1973; Dijken and Harder, 1975; Babel, 1979).

The theory of fermentation has failed to provide reliable methods for obtaining the theoretically calculated maximum yield values. The introduction of the maintenance coefficient into the fermentation theory by Pirt (1965) has led to certain progress because of the assumed proportionality between the specific growth rate constant and the yield of cell mass derived from this concept:

$$(10) \quad \frac{1}{Y} = \frac{1}{Y_{max}} + \frac{m}{\mu}$$

A closer consideration of this concept gives rise to the assumption that the maintenance coefficient as defined by Pirt is the sum of a real maintenance coefficient and a coefficient which takes energy dissipation into account. Thus, the maintenance coefficient is influenced by changing environmental conditions which cause an increased specific heat production and material consumption coefficients (Ringpfeil, 1982a).

Thus the approximation of yield values to the theoretically calculated maximum values is still a time-consuming empirical effort comparable with that of the approximation of the specific growth rate constant to its maximum.

Compared with the intensive work on the theoretical background of fermentation, almost nothing has been done to elucidate the natural behaviour of the recovery processes. Recovery of SCP is almost exclusively covered by physico-chemical theories, which regard the cell as a biologically inactive particle of more or less colloidal behaviour. Therefore, the methods applied are based on the unit operation concept known from the chemical industry. Necessary progress can possibly be obtained only if in the theoretical considerations the biologically active nature of the cell coming out of the fermenter is taken into account. This refers to the specific activities of the cell walls and the ability of the cells to metabolize nutrients before procedures such as heating occur, as well as to the property of cells to serve as a nutrient for contaminating organisms during storage in humid surroundings before and after destroying their biological activity.

TECHNOLOGICAL BASIS OF SCP PRODUCTION AND RECOVERY

A broad extension of the raw materials used has been observed during the last three decades. Today all basic organic materials are used for SCP synthesis (Fig. 2; Ringpfeil, 1979), with the use of organic wastes to be expected in the near future. The costs of the carbon sources for fermentation rise with each step involved in processing the basic raw materials. On the other hand, the costs of recovering SCP diminish with the number of purification steps of the carbon source. Natural gas plays a special role in being a relatively pure carbon source.

Rising costs of carbon sources have caused a search for cheaper raw materials, preferably in organic wastes. The use of wastes as raw material for SCP production is nothing unusual; sulphite liquor, thin stillages, and whey have been used for more than 50 years (Ringpfeil *et al.*, 1981a). A search for further sources suitable for industrial conversion has been initiated. They were found in the organic wastes of the huge animal production units. Swiss agricultural experts have estimated that about 25% of the energy (as reduced carbon) of total primary plant production is hidden in manure (Fiechter, 1980). In the G.D.R., for instance, several pig production factories with an output of more than 200 000 units/year, discharge approximately 3 000 m^3 of liquid waste per day with an organic load of about 10 000 to 60 000 mg O_2/litre (Ringpfeil *et al.*, 1981a). These wastes are heterogeneous in their chemical composition, but could be qualified as 'physiological', so that fodder production by microbial conversion seems to be possible as far as medical considerations are concerned. This field of organic reserves is only now being 'tapped.' The obvious advantage of combining sewage purification with the production of valuable goods will lead to further efforts in making accessible other sources of waste materials that are suitable for recycling as fodder by microbial means.

Another source of raw materials will be opened up by the use of agricultural products which may be partially converted into protein by semisolid fermentation, taking advantage of the fact that unconverted material will either retain a certain nutritional value as carbohydrate or fat or 'will not cause harm' for being inert (Senez, 1983; Hesseltine, 1972 and 1977).

For conversion into cell mass the oxygen content of the carbon source plays primarily a negative role because it diminishes the energy content per unit weight of the molecule, and thus increases the necessary amount of raw material per unit weight of cell mass.

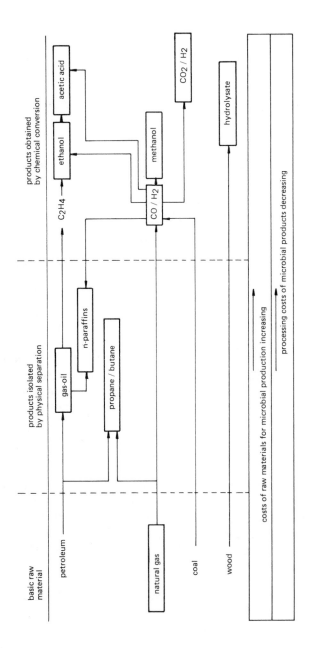

Fig. 2.   Raw materials for the production of microbial feedstuffs.

The type of microorganism used to convert a chosen substrate depends on several circumstances. Firstly, the ability of the microorganism to ferment the substrate. Secondly, if there is a choice between bacteria, yeast, and fungi, yeasts tend to be perferred for historical reasons, and simply expressed in the widespread opinion that 'bacteria cause sickness, and fungi possibly too, and that yeasts have been used in baking bread and cake for ages'. More cogent reasons for preferring yeasts are their higher protein content compared with fungi, lower nucleic acid content and better properties for recovery in comparison with bacteria (e.g. larger cells and no slime capsule; Ringpfeil 1982a). But despite these reasons, bacteria should grow in importance because of their ability to convert significantly more organic compounds than yeasts or fungi. Fungi cannot be disregarded because of their ability to excrete enzymes, which are capable of degrading polymeric substances into fermentable molecules, e.g. cellulose or starch into glucose. From the present point of view, several preferential combinations of substrates and microorganisms are accepted (Table II). It must be taken into account that most fermentations to SCP are carried out under conditions which allow faster growing contaminating organisms to outgrow the production organism. It has been observed that fungi are not able to withstand air-borne yeasts in the continuous fermentation of sulphite liquor (Rieche, 1954). In unprotected continuous methanol fermentations, yeasts will be displaced by acidophilic bacteria (Schneider and Meyer, 1980). On the other hand, most bacteria cannot withstand yeasts when fermented in acidophilic carbonhydrate-containing materials.

Generally, SCP production is carried out without the artificial help of growth stimulators. But it must be taken into account that these fermentations often occur under conditions which allow certain types of 'accompanying organisms' to grow to a limited extent beside the production culture. If these organisms are excluded by sterile fermentation

TABLE II
Preferentially used microorganism/substrate combinations

Microorganism	Substrate
Bacteria	Methane, methanol, organic wastes
Yeasts	Methanol, ethanol, paraffins, sugars, sugar-containing wastes
Fungi	Starch, cellulose, sugars

techniques the production culture will suddenly fail to grow without stimulators (Brendler *et al.*, 1982). It may thus be assumed that the accompanying organisms play an important role in providing the production culture with necessary growth factors. Mixed cultures for SCP production should not be condemned as has sometimes been done, but rather studied with greater intensity in order to obtain further knowledge about the risk they may entail, as well as their economical advantage. Heterogeneously composed substrates such as manure cannot be converted successfully without the aid of mixed cultures (Ringpfeil *et al.*, 1981b). Technical devices for providing the fermentation with molecular oxygen have been known since the beginning of the technical exploitation of the aerobic growth of microorganisms (Delbrück, 1915). Because of the low content of fermentable substrate of the available liquid raw materials, the oxygen transfer capacity of these devices remained low (Ringpfeil *et al.*, 1981a). The use of concentrated substrates gave rise to the development for more efficient mass transfer devices. The IZ jet stream reactor (Jagusch, 1968) represents those high-performance fermenters that are built to a volume of several thousand cubic meters, and form the basis for modern SCP industries.

The most frequently applied chemostat type of continuous cultivation possesses a certain autoregulatory ability to keep cell mass concentration constant. Necessary artificial aids are the regulation of *p*H, temperature, medium inflow, mechanical agitation and sometimes foam development, which can be achieved by simple regulative cycles. With the development of computerization, some efforts have been made on its introduction into SCP production systems. The principle reason for its use is to equalize disturbances of fermentation processes caused by variations in the chemical composition of raw materials, the supply of materials and energies to the fermenter as inevitably occurs in practice. Finding the new optimum or the most efficient way to reach the desired state of fermentation again, could be realized more easily through computers connected to the fermenter. This may lead to a reduction in costs by 5% (Albrecht *et al.*, 1981).

Recovery is performed on the basis of the mechanical and thermal removal of liquid from SCP (Ringpfeil, 1979). The concentration of cell mass is accomplished step by step depending on its physical properties. Rising cell mass concentration in fermenter outflow due to a more intense oxygen mass transfer in the fermentation step leads to projects designed to eliminate the mechanical separation and connect directly

fermentation with evaporation. The low concentration grades reached by mechanical separation due to the small size of cells and more or less voluminous slime capsules in bacterial suspensions promote this development. But it may be expected that the mechanical separation will increase in importance again when new results about the agglomeration of cells are presented. Among the mechanical separation methods, flotation should not be overlooked. Methods such as the two-stage flotation applied recently in sewage sludge removal (Zlokarnik, personal communication) should be watched carefully.

TRENDS FOR ECONOMIZATION OF SCP PRODUCTION

Economization of SCP production is necessary in order to gain advantages over the agricultural path of protein production (Ringpfeil, 1979 and 1982b). The development of industrial protein production is only justifiable if it becomes the more efficient method. A simple way of assessing efficiency is by estimating the costs of the production of the target product. But there are certain cases where costs do not show the real expenditure for the process. Detailed analysis seems to be necessary to determine those parts of the production which contribute most to the decrease in expenditure. Such analysis give specific results for any local conditions. Nevertheless, some factors of general importance can be deduced. They are

- producing other products besides protein during the production process,
- simplifying downstream processes,
- augmenting velocity and yield of fermentation,
- choice of substrate (if possible).

Other Products

The use of gas oil as a raw material causes several specific problems in the SCP process due to its heterogeneous nature. The remaining gas oil (up to 90% of the amount used) must be separated from the cell mass; this is commonly performed in two operations: removal of the main proportion by mechanical separation from the aqueous cell mass leading to deparaffinized gas oil, and removal of the residual amount by solvent extraction from the cells leading to a residual after distillation of the solvent.

The deparaffinized gas oil possesses a low freezing point. It can be used to advantage in engines under cold wintery conditions, as special oil in transformers (Hieke *et al.*, 1982) and in similar applications.

The extraction of the residual hydrocarbons located in the surface and inside the cells can be performed quantitatively only when the protein-lipid complexes inside the cell are broken by the action of polar solvents before or during the soaking of cells with apolaric solvents (Ringpfeil, 1982b; Biedermann *et al.*, 1981). The residuals obtained after removing the solvents from the mixture contain, in addition to gas oil, lipophilic constituents of the microbial cell, e.g. phosphatides, fatty acids, glycerol, and minor compounds such as ergosterols and ubiquinones (Ringpfeil, 1979; Voigt *et al.*, 1979; Müller and Voigt, 1982). Many applications may be found for the residual, the so-called biolipid extract as well as for its single compounds (Tables III and IV).

The SCP production process on gas oil delivers at least three final products: biomass, deparaffinized gas oil, and biolipid extract (see

TABLE III
Application of biolipid extract

Application field	Application concentration (%)	Technical use
Fluorite flotation	0.04	Collector
Paint production	2	Preventor of pigment deposition
Oil for heating plants	0.05	Promotor of combustion
Road-making	1	Improver of adhesion between bitumen and grit layer
Production of fertilizers	0.03	Anti-jaw agent
Production of plant-protection agents	5	Emulsifier, demulsifier
Building industry	30	Separation oil in moulding boxes
Geological drilling	5	Surface active agent
Petroleum processing	10	Improver of bitumen properties
Agriculture	10	Improver of soil
Environment protection	10	Emulsifier of petroleum
Biotechnical processes	0.1	Improver of oxygen transfer and separation of yeast/oil mixtures

TABLE IV
Application of biolipid extract fractions

Fraction	Field of application
Crude phosphatides	Road-making, production of plant-protection agents, production of paint
Pure phosphatides	Substitution of plant lecithins in different branches of industry
Fatty acids	Fluorite production, paint industry, waste water treatment, fungicide production
Ergosterol	Raw material for production of Vitamin D_2 and steroid hormones
Ubiquinone	Pharmacology, growth regulator of Mycobacteria

Fig. 1). The analogous composition of microbial cells produced from different raw materials always permits the production of biololipids as well as hydrophilic products such as nucleic acids or polysaccharides without decreasing the absolute amount of protein destined for food and feed. General knowledge about fermentation makes it possible to vary the amount of non-proteinaceous constituents of cells according to market considerations (Sattler *et al.*, 1980).

Until now there have been no economic suggestions to recover the biggest by-product of SCP production: heat. Any scheme for SCP production from gas oil shows the tremendous amount of heat wasted by (expensive) cooling in the fermentation step or released by hot gases and steam (Ringpfeil, 1979). Most of the heat is produced at a temperature level permitting its recovery only if considerable amounts of external energy are used, which is economically problematic and rarely useful locally. Raising the fermentation temperature would therefore be possible, but the knowledge about achieving maximum yield values using thermophilic microorganisms has remained poor.

Downstream Processes

The simplification of downstream processes starts with the rise of cell concentration in the outflow of the fermenter. Besides increasing cell concentration by applying high-performance for oxygen transfer, gas oil conversion shows an example for further improvement. The walls of

yeast cells are partially hydrophobic. Thus, yeast cells and the remaining oil combine to form a floc of a density lower than the aqueous phase. The surfacing flocs may be separated from the main body of water by simple skimming.

The separation of the oil/cell mixture is commonly performed by separating machines which demand skilful handling. A more economic solution is promised by the heterocoalescence effect. The oil/cell mixture flows along the surface of a granulate made of apolaric organic polymers. The affinity of the oil to these surfaces is greater than that to the yeast surface. A rearrangement of molecular forces occurs and the oil separates from the cells under the actions of surface-active agents and raised temperature (Konieczny and Pickert, personal communication).

Present knowledge about the properties of aqueous suspensions of microbial cells and of the cells themselves is rather poor. More detailed investigations are necessary in order to find further simplifications of downstream processes.

Velocity and Yield of Fermentation

Increasing the velocity of fermentation was tried for a long time without any real knowledge about the natural limits of the productivity of cell growth processes. These limits could be determined by a modified productivity equation (Ringpfeil, 1982b):

$$(11) \qquad \dot{x}_{max} = \mu_{max} \cdot x_{max}$$

This equation seems to contradict the Monod equation because, following it, x_{max} could not be obtained at μ_{max}. But it has to be taken into account that k_s in the Monod equation (Equation 7) by no means shows the quality of a material constant. Like the maintenance coefficient, it appears as a sum which may be composed of a real k_s value and a value caused by ineffective fermentation conditions. Evidence for such a hypothesis is provided by growth experiments showing k_s to be equal to zero or almost zero (Ringpfeil, 1980).

For calculating x_{max}, values for μ_{max} were derived from experiments realizing the highest possible growth rate constant at lower x than x_{max} values. These μ_{max} values were combined with x_{max} values derived from rheological studies of the fermentation broth carried out by adding dry cells (Ringpfeil, 1980). Although methodology could not

exclude some uncertainties, surprisingly high values of maximum productivity could be derived for certain cell/substrate combinations (Table V; Ringpfeil, 1979). Experimental verification shows the usefulness of these considerations. Moreover, the results led to the conclusion that the maximum values of productivity should be calculated generally because their difference from the experimentally obtained ones gives a measure of the possibility of intensification of the investigated processes (Ringpfeil, 1982b).

TABLE V

Attainable productivities of microbial fermentation of cell mass

Microorganism/substrate combination	Attainable productivity $[kg/m^3h]$
Yeast/gas oil	10
Yeast/n-paraffins	25
Yeast/ethanol	50
Yeast/carbohydrates	50
Bacteria/methanol	25
Bacteria/methane	25

Besides a careful determination of nutrient concentrations, pH, temperature, and other environmental conditions, an oxygen transfer velocity must be guaranteed which is adequate to the calculated maximum productivity values. For economic reasons it is essential to realize this oxygen transfer velocity at minimum levels of energy and gas inputs. Fundamental considerations about the most favourable oxygen transfer system were necessary, and were based on the kinetics of the participating reactions and the hydrodynamic properties of the applied technical system (Ringpfeil, 1981b). In contrast to the usual assumptions two reactions were considered: the first-order reaction of oxygen exhaustion of gas bubbles by transfer and consumption

$$(12) \quad \dot{c} = K_L a H P_{O_2}$$

and the autocatalytic cell growth reaction (Equation 3). Consequently, for continuous processing these two reactions require two contradictory hydrodynamic systems: a plug-flow system for oxygen exhaustion and a mixed vessel for cell growth. The so-called Cycle Tube Cyclone Reactor

(C.T.C.R.; Fig. 3; Ringpfeil, 1981; Ringpfeil and Stephan, 1979; Ringpfeil *et al.,* 1981; Stephan, 1980) consists essentially of a tube which allows a vertical down-stream of a mixture of gas and liquid. After the medium had left the tube, a cyclone separates the exhausted gas from the liquid. Over a length of 10m, the mixture flows around for 30 sec, transferring up to 70% of the oxygen from the gas into the liquid (provided a catalyst converting molecular oxygen is present in right amounts). The liquid is driven by a circulating pump, and newly mixed with air, flows again into the tube, which is surrounded by a cooling jacket. Inside the tube, sieve plates are located in a geometric downward-directed progression permitting increasing mechanical energy input with decreasing oxygen tension in the gas (Ringpfeil, 1980). In this

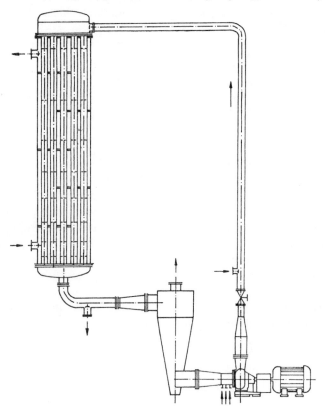

Fig. 3. Scheme of the Cycle Tube Cyclone Reactor (C.T.C.R.).

manner a minimum of energy expenditure is obtained. In the tube, plug flow is achieved during one cycle which is performed within seconds. The cyclone and the pump permit sufficient mixing for the autocatalytic growth taking place in hours. Thus at least both contradictory characteristics of flow may be obtained in one device regarding the different time course of the reactions involved. This reactor construction seems to be a model for an ideal hydrodynamic system for aerobic cell growth, with the attained values of oxygen transfer rate and productivity suppling the evidence (Table VI; Ringpfeil *et al.*, 1981a).

TABLE VI
Capacity of high-performance fermenters

Reactor	Substrate/	V_F (m^3)	$\dot{x}$ (g/kg h)	$\dot{c}$ (g/kg h)	a_x^p (Wh/g)
IZ Jet Stream	Sucrose/Yeast	0.3	12–16	11–15	0.24–0.28
Reactor	Ethanol/Yeast	0.3	6–14	9–23	0.5–0.7
Cycle Tube	Sucrose/Yeast	0.05	15.1	32.1	1.75
Cyclone Reactor	Ethanol/Yeast	0.05	34.0	55.5	1.1

The technical construction of the IZ jet stream reactor comes closest to the model, if compared with other systems. High values of oxygen transfer rate and productivity could be observed (Table VI). Fermenters in a two-storey construction with volumes of up to 2200m³ have now been put on stream (Jagusch and Schreck, 1980).

Other means of enhancing the oxygen transfer rate have been investigated. Most promising is the addition of surface-active agents (Ringpfeil, 1981b). These enlarge the surface between the gas and the liquid and decrease the permeability of the surface layer to the gas molecules. Substances have been found which empirically lead to an overall increase of the oxygen transfer rate. One of these substances is composed of ethylene/propylene glycols, introduced as Ferman in commercial fermentation (Klappach and Otto, 1980).

Other methods for increasing the oxygen transfer rate are higher pressure and/or higher oxygen tension. Productivity has been found to increase in a linear progression with increasing partial pressure of oxygen (Ringpfeil *et al.*, 1978). This method has become especially useful in fermentations of natural gas where the partial pressures of

methane and oxygen mutually diminish their driving force (Wendtland, unpublished results).

Thus, productivity could be raised by one order of magnitude without any considerable increase in specific energy consumption at high productivity values (Ringpfeil, 1980). This development has accelerated the design of large capacities for SCP production without the need for large fermentation volumes. But it does not seem that the development of high-performance fermenters with large output has come to an end. Further development should still make further economization possible.

Yield has attracted scientific effort since pure raw materials of higher refinement such as methanol have been introduced into SCP production.

One of the most exciting examples is the Dynamic Process Control Concept (D.P.C.C.) (Heinritz *et al.*, 1984). Its basis is the well-known occurrence of different cell states during the growth cycle (Meyenburg, 1969; Streiblova, 1980; Dawson, 1972; Bley *et al.*, 1980). Yeasts show remarkably different demands for substrate during their different states (Ringpfeil, 1982 and 1982b; Bley *et al.*, 1980; Heinritz *et al.*, 1981 and 1983). A uniform substrate concentration causes losses by respiration of the excess substrate of those organisms having a lower substrate demand. By synchronizing growth and supply of substrate according to the demand during the different phases, overall substrate consumption could be decreased in the range of 10 to 15% (Table VII; Ringpfeil, 1982b). Moreover, some further properties of the fermentation process and the product could be changed advantageously (Table VII; Bley *et al.*, 1978). The method is far from exhausted, and several advantages could be still expected following more detailed study.

Another example is mixed-substrate fermentation follwing the Auxiliary Substrate Concept (A.S.C.) (Babel, 1979). It has, as its basis, the different distribution of carbon and energy available for cell growth in substrate molecules (Ringpfeil, 1982b; Babel, 1980; Heinritz *et al.*, 1982). Microorganisms which are able to utilize more than one substrate simultaneously can make use of the balanced carbon and energy content introduced by adding two or more substrates in a precalculated manner. In this case a 10% increase in yield could be obtained compared with the yields received by fermentation of single compounds (Fig. 4; Richter and Ondruschka, 1980).

Yield is often influenced by the fact that microbial populations react an short-time periodic changes of environmental conditions with an

TABLE VII
Dynamic processing in the fermentation system gas oil/yeast-comparison to steady-state fermentation (period of changing substrate feeding rate of 12 hours, synchronisation by pH-value shocks and changes of substrate feeding rate)

Process control concept	Average gas oil feeding rate [g/kgh]	D [h]	$\dot{x}$ [g/kgh]	$Y_{s/x}$ [g/g]	$Y_{c/x}$ [g/g]	$Y_{Q/x}$ [kJ/g]
Steady state	40	0.2	2.64	1.2	2.4	8.5
Dynamic state	27	0.2	2.6	1.05	2.0	6.8

Process control concept	Freezing point of de-paraffinized gas oil [°C]	Lipid content of cell mass [%]	Carbohydrate content of cell mass [%]	Crude protein content of cell mass [%]	Ash content of cell mass [%]
Steady state	− 28	22	15	57	6
Dynamic state	− 40	17	12	64	6

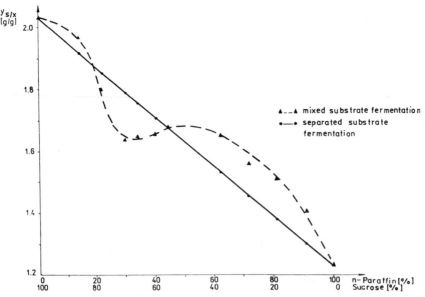

Fig. 4. Mixed-substrate fermentation following the Auxiliary Substrate Concept in the fermentation system paraffin/sucrose/yeast.

increased heat and carbon substrate dissipation. Such changes occur preferably in recirculation reactors in which microorganisms are passing zones of different concentration levels in the broth. Simulation of this behaviour shows a reaction of the culture defined as hysteresis. To overcome this an analysis of the hydrodynamics of the fermenter and the technical arrangements for maintaining homogeneity and/or choice of optimum running parameters is necessary (Ringpfeil, 1982a; Heinritz, 1978; Glombitza and Heinritz, 1979; Heinritz and Bley, 1979).

Substrate Choice

Another fundamental method of economizing SCP production is based on the introduction of waste materials as carbon (as well as nitrogen and phosphorus) sources apart from the use of carbohydrate wastes. These raw materials necessitate expenditure for their conversion into harmless compounds such as CO_2 and water. Such wastes are inexpensive carbon sources for SCP production possessing the added benefit of being recycled. Their disadvantages is their heterogeneous chemical composition. It seems to be difficult to determine the residual compounds after fermentation. It is also difficult to maintain a determinable composition of organisms because the heterogeneous nature of the chemical composition necessarily needs a mixture of different microorganisms. A solution can be found by introducing new technical elements into the SCP process. One such is achieved by adding a pure carbon source to

TABLE VIII

Chemical composition of SCP, biosludge and biomass from feedlot waste effluents fermented with addition of methanol

System	Crude protein (% dm) (N × 6.25)	Nucleic acids total (% dm)	Lysine (g/kg dm)	Ash (% dm)
MB 58/MeOH/Feedlot waste effluent	72	7 −10.5	32−40	6 − 7.5
MB 58/MeOH	75	9.6−11.0	36	4.5− 5.5
Candida utilis/EtOH	57	7.8− 9.8	32	5 − 8
Biosludge from an aerobic activated sludge process	40−60	6 − 9	10−25	15 −30

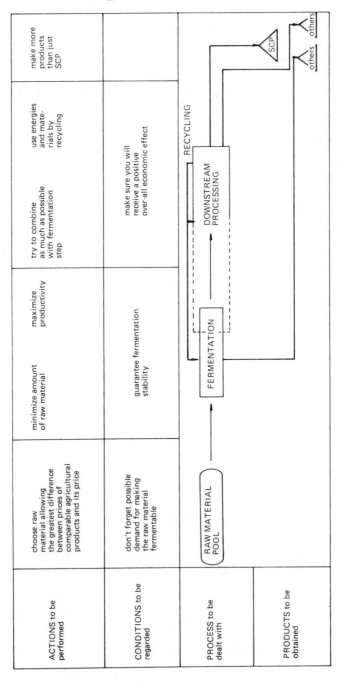

Fig. 5. Mode of operation in developing SCP processes.

the original sewage. In the case of manure the C:N:P ratio is deficient in carbon for conversion into microbial cells. Adding a carbon source such as methanol or glucose would correct the C:N:P ratio to the right proportions. Lower C:N:P-values occur in the fermented liquid (Ringpfeil et al., 1981). The chemical composition of the resulting cells does not deviate from that obtained from a pure carbon source (Table VII). A microbial population containing more than 80% of a single organism is observed, but it has to be noted that the residual 20% contain at least all the species found in the original sewage sludge (Ringpfeil et al., 1981b). Another element is the introduction of a high-temperature short-time alkalinization of the grown cells to destroy their structure and safely kill them (Ringpfeil et al., 1981a, c). Thus the prerequisites for a safe application for fodder purposes have been obtained. Nevertheless, testing has to be carried out in accordance with PAG and IUPAC guidelines in order to adapt them if necessary to the specialities of these raw materials, processes and products (Ringpfeil, 1982c).

In conclusion a simple scheme can be composed to show the main operations which could be performed in order to increase the economy of SCP production (Fig. 5; Pimentel et al., 1973). Differentiated application is necessary to take into account local conditions and achieve optimum results.

FURTHER ECONOMIC CONSIDERATIONS CONCERNING
SCP PRODUCTION

A comparison with agricultural alternatives of protein production may be attempted by considering the energetic balance. Basic methods for such calculations are examples of the quantitative determination of energy consumption in corn production (Pimentel et al., 1973). It may be shown that SCP production on organic raw materials is not energetically self-sufficient as is agriculture (Böhme et al., 1978). But on a protein-based calculation it can be concluded that even modern agriculture tends towards dependency on stored energy. This tendency is caused by the rising demand for energy-consuming auxiliaries and the continuing demand of considerable working power. The method of calculation applied for corn production (Pimentel et al., 1973) contains a serious error. Energy demand for working power is calculated only as physiological energy demand, yet a worker will consume a far greater amount of objective energy. It is assumed that the overall energy consumption for non-productive purposes divided by the number of productive

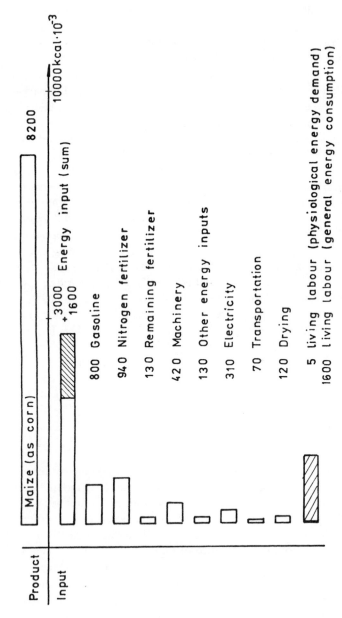

Fig. 6. Balance of industrial production of corn (Pimentel *et al.*, 1973).

employees will give a rough figure of the total energy a worker needs (Ringpfeil and Koriath, 1984). If that value is introduced, the energy demand for corn production changes significantly (Fig. 6). Based on that — more realistic — picture of the energy demand, the advantages of SCP production become more obvious due to its industrial organization of production, permitting a high productivity of labour. Industrial SCP production based on fossil raw materials becomes at least energetically comparable to agricultural protein production.

SCP production gains specific advantages if it is linked to other production. SCP from gas oil will be produced more profitable if it can be incorporated into a petrochemical complex. Products of the petro-chemical industry (gas oil) are raw materials for the biological process and some products of which (deparaffinized gas oil, lipids) are raw materials for new petrochemical productions. SCP production from manure has specific advantages if it is incorporated into animal produc-tion units (Fig. 7; Ringpfeil and Kehr, 1980). By introducing a combined sewage treatment-biomass production process (S.T.B.P.) the purchase of feed ingredients could be reduced ($B < B'$) or the separate production of protein could be eliminated ($B'' = O$), respectively. Due to the use of carbon sources from the manure raw material, demand is lower than in the case of separate protein production ($A < A'$). The fermentation devices in an S.T.B.P. plant have a considerably lower working volume than those in the conventional sewage treatment plant ($C < C''$). The effluent from the combined plant is lower than that for the conventional one, which is due to more intensive exhaustion of ingredients ($D < D'$).

CONCLUSIONS

SCP production is a new and promising form of productive force which advantageously combines both biological specificity and the ability to convert simple raw materials into complicated, biologically active matter. Raw materials may be different and heterogeneous. Fermenta-tion has been developed to a level approaching the intensity of chemical processes. The colloidal structure of SCP and its ability to retain water influence the recovery of biomass unfavourably. The heterogeneous composition of the resulting biomass makes it possible to separate certain compounds more efficiently, and this may be applicable for industrial purposes rather than for nutrition. The isolation of protein is possible and permits the production of pure proteins, which is especially

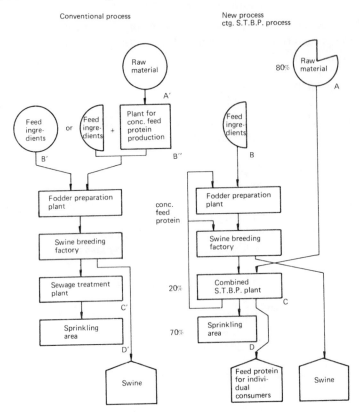

Fig. 7. Sewage treatment scheme for large swine breading systems.

useful for human nutrition. Medical and biological research and the broad testing of different SCP products have convincingly proved their safety and value for use as food and feed.

At present, SCP production mainly complements agricultural protein production. The advantages offered by the industrial organization of SCP production makes for a high productivity of labour and is economically advantageous, despite the use of organic raw materials. Different branches of industry such as the chemical, pulp and paper, sugar, food and alcohol industries, as well as agriculture, can introduce SCP production into their production schemes using their products, by-products, and wastes as raw materials. The often-voiced opinion that fossil organic raw materials will soon be exhausted and that SCP production should not be based on them is without substance, since

SCP production can make use of primary organic raw materials. Certain raw materials are far from being in short supply for food and feed when the conversion of non-edible cellulosics into protein is considered. Thus, at least the availability of raw materials of fossil origin does not limit SCP production. Its feasability is dependent only on the economic accessibility of raw materials and their economic processing. These problems can be solved by the scientific development that is presently under way in several places of the world.

APPENDIX: SYMBOLS USED IN TEXT

Y, Y_{max} Yield coefficients, maximum yield coefficient

$Y_{x/s}$ Biomass yield in g of dry-weight cell mass per g of substrate

$Y_{x/c}$ Biomass yield in g of dry-weight cell mass per g of oxygen

Y_{x/NH_4^+} Biomass yield in g of dry-weight cell mass per g of ammonium ions

$Y_{x/PO_4^{3-}}$ Biomass yield in g of dry-weight cell mass per g of phosphate ions

$Y_{CO_2/x}$ Specific carbon dioxide production in g of carbon dioxide per g of dry-weight cell mass

$Y_{Q/x}$ Specific heat production in kilojoule per g of dry-weight cell mass

$Y_{s/x}$ Specific substrate consumption in g of substrate per g of dry-weight cell mass

$Y_{c/x}$ Specific oxygen consumption in g of oxygen per g of dry-weight cell mass

x Biomass concentration in g of dry-weight cell mass per kg of fermentation medium

x_o Initial biomass concentration in g of dry-weight cell mass per kg of fermentation medium

s Substrate concentration in g of substrate per kg of fermentation medium

k_s Substrate concentration at $\mu_{max}/2$ in g of substrate per kg of fermentation medium

$\dot{x}, dx/dt$ Biomass productivity in g of dry-weight cell mass per kg of fermentation medium and hour

$\dot{s}$ Substrate feeding rate in g of substrate per kg of fermentation medium and hour

$\dot{c}$ Oxygen transfer rate in g of oxygen per kg of fermentation medium and hour

$\dot{x}_{max}$	Maximum biomass productivity
x_{max}	Maximum biomass concentration
μ	Specific growth rate in reciprocal hours
μ_{max}	Maximum specific growth rate constant
$K_L a$	Specific oxygen transfer rate constant
P_{O_2}	Dissolved oxygen tension
m	Manintenance coefficient in g of substrate per g of dry-weight cell mass and hour
α_x^P	Specific energy input in Wh per g of dry-weight cell mass
D	Flow rate
V_F	Fermenter volume
T	Doubling time of microorganisms
t	Time
H	Henry constant

REFERENCES

Albrecht, Ch., Pöhland, D., Prause, M. and Ringpfeil, M. (1981). 'Optimal Control of SCP Fermentation Processes.' *Proc. III. Int. Conf. Computer Appl. in Fermentation*, pp. 191–197. Manchester (U.K.).

Babel, W. (1979). *Z. Allg. Mikrobiol.* **19**, 671–677.

Babel, W. (1980). *Acta Biotechnologica* **0**, 61.

Biedermann, W., Heinze, G., and Schmidt, J. (1981). *Zu Fragen des Mechanismus und der Kinetik der Extraktion von Kohlenwasserstoffen aus ED-Hefen. Sitzungsber.* Akad. Wiss. DDR 16 N/1980; Akademie Verlag Berlin.

Blanc, B. (1978). Alimenta **17**, 155.

Bley, T., Glombitza, F., Heinritz, B. and Rogge, G. (1978). *Patentschrift DD WP 140* 150.

Bley, T., Heinritz, B., Steudel, A., Stichel, E., Glombitza, F. and Babel, W. (1980) *Z. Allg. Mikrobiol.* **20**, 283–286.

Böhme, H., Keil, G., Philipp, B. and Ringpfeil, M. (1978). *Bedeutung und Einfluss der Chemie auf die Entwicklung der industriemässigen Produktion der Landwirtschaft.* Sitzungsber. Akad. Wiss. DDR 19 N/1978.

Brendler, W., Bauch, J., Lübbert, G. A., Wünsche, L., Hädlich, R., and Shdannikowa, E. N. (1982). 'Special Aspects of Nonsterile Production of Fodder Yeast on the Basis of Petroleum Hydrocarbons.' *Proc. 'Leipziger Biotechnologie-Symposium' Leipzig.*

Dawson, P. S. S. (1972). *J. Appl. Chem. Biotechnol.* **22**, 79–103.

Delbrück, M. (1915). *Z. Spiritusindustrie* **38**, 121.

Dijken, J. P. and Harder, W. (1975). Biotechnol. Bioeng. **17**, 15.

Fiechter, A. (1980). *Proc. II. Rothenburger Symposium.* Bad Karlshafen.

Glombitza, F. and Heinritz, B. (1979). *Z. Allg. Mikrobiol.* **19**, 171.

Hamer, G., Harrison, D. E. F., Harwood, J. H. and Topiwala, H. H. (1975). In *Single Cell Technology II* (S. R. Tannenbaum and D. I. C. Wang, eds.). MIT Press, Cambridge.

94   M. Ringpfeil and B. Heinritz

Heinritz, B. (1978). Dissertation. Akad. Wiss. DDR, Leipzig.
Heinrtiz, B. and Bley, T. (1979). Z. Allg. Mikrobiol. 19, 247–252.
Heinritz, B., Bley, T., Rogge, G. and Ringpfeil, M. (1984). Acta Biotechnologica 4, 275.
Heinritz, B., Stichel, E., Bley, T., Rogge, G. and Glombitza, F. (1981). Z. Allg. Mikrobiol. 21, 581–586.
Heinritz, B., Stichel, E., Rogge, G., Bley, T. and Glombitza, F. (1982). Z. Allg. Mikrobiol. 22, 535–544.
Heinritz, B., Rogge, G., Stichel, E. and Bley, T. (1983). Acta Biotechnologia 3, 125.
Hesseltine, C. W. (1972). Biotechnol. Bioeng. 14, 517–532.
Hesseltine, C. W. (1977). Process Biochem. 12, 24–32.
Hieke, W., Kockert, M., Gentsch, H., Bauch, J. and Bohlmann, D. (1982). 'Einsatz von paraffinhaltigen Erdöldestillaten für die Herstellung von Transformatorölgrundkomponenten.' Chem. Techn. 34(8), 418–419.
IUPAC (1974). Technical Report No. 12. Appl. Chem. Divison.
IUPAC (1978). Technical Report. Final Draft on Proposed Guidelines for Testing of Single Cell Protein. Appl. Chem. Divison.
Jagusch, L. and Püschel, S. (1968). MWT 18, 160–331.
Jagusch, L. and Schreck, R. (1980). 'Techn.-ökon. Untersuchungen am IZ-Strahlfermentor unter Druck.' Proc. II. Symp. der soz. Länder über Biotechnologie, Leipzig.
Jones, R. W. (1951). CEP 47, 46.
Klappach, G. and Otto, G. (1980). 'Zum Einfluss von Tensiden auf den O_2 -Übergang und die Produktivität der SCP-Bildung.' Proc. II. Symp. der soz. Länder über Biotechnologie, Leipzig.
Lehninger, A. L. (1970). Bioenergetik. Thieme-Verlag, Stuttgart.
Liepe, F., Jagusch, L., Stephan, W. and Ringpfeil, M. (1978). 'The Present State and the Perspective of Development of Industrial High-Performance Reactors for Aerobic Fermentation.' Proc. I. Europ. Kongress Biotechnologie, Interlaken.
Malek, I. (1958). Continuous Cultivation of Microorganisms – A Symposium. Pub. House of the Czechoslov. Acad. Sci., Prague.
Malek, I. (1964). Continuous Cultivation of Microorganisms II. Pub. House of the Czechoslov. Acad. Sci., Prague.
Meyenburg, K. V. (1969). Arch. Microbiol. 66, 289.
Mitchell, P. (1961). Nature 191, 144.
Monod, J. (1949). Ann. Rev. Microbiol. 3, 371.
Monod, J. (1950). Ann. Inst. Past. 79, 390.
Müller, H. and Voigt, B. (1982). Acta Biotechnologica 2, 155–160.
Norris, J. R. (1968). Advancement Science 25, 143–150.
Payne, W. J. (1970). Ann. Rev. Microbiol. 24, 17.
Pimentel, D., Hurd, L. E., Bellotti, A. C., Forster, M. I., Oka, I. N., Sholes, O. D. and Whitman, R. I. (1973). Science 182, 443.
Pirt, S. J. (1965). Proc. Royal Soc. B 163, 224.
Pokrowskij, A. A. (1972). Mediko-Biologitsheskiya Issledovaniya Uglevodorodnych Droshey. Izdat. Nauka, Moscva.
Pokrowskij, A. A., Laube, W., Aksjuk, J. N., Henk, G., Uschakowa, T. M. and Herrmann, U. (1978). Chem. Techn. 30, 368–371.
Protein Advisory Group (1967). Documents on Single Cell Protein issued by PAG on

15.3.1967: PAG Guidelines No. 6 for preclinical testing of noval sources of protein, 1972; PAG guideline No. 12 on the production of single cell protein for human consumption, 1972; PAG guideline No. 15 on nutritional and safety aspects of noval protein sources for animal feeding, 1974.

Protein Advisory Group/UNU (1982). Guidelines No. 6 for preclinical testing of noval sources of food; PAG/UNU guidelines No. 12 on the production of single cell protein for human consumption; PAG/UNU guideline No. 15 on nutritional and safety aspects of protein sources for animal feeding, June 1982.

Rieche, A. (1954). *Wiss. Ann. Dtsch, Akad. Wiss.* **3**, 705–728.

Richter, K. and Ondruschka, J. (1980). 'Mischsubstratfermentation in System Sacch./ Paraff./Hefe.' *Proc. II. Symp. der soz. Länder über Biotechnologie,* Leipzig.

Ringpfeil, M. (1978). 'Fragen der Mikrobiologie und wissenschaftliche Grundlagen der mikrobiologischen Industrie.' *Wiss. Session Neubrandenburg 1975. Abh. Akad. Wiss. DDR. Abt. Math. Naturw., Techn. Nr.* l; Akademie Verlag Berlin.

Ringpfeil, M. (1979). 'Microbial Processing – A Technical and Economic Appraisal. *Proc. X. World Petroleum Congress* **4**, 377–385. Heyden & Son, London.

Ringpfeil, M. (1980). 'Biotechnologie auf dem Weg zur Wissenschaft.' *Proc. II. Symp. der soz. Länder über Biotechnologie,* Leipzig.

Ringpfeil, M. (1981a). 'Fermenter design for SCP production.' In *'VIth Int. Fermentation Symp., London 1980'* (M. Moo-Young, ed.) p. 741; Pergamon Press, Toronto.

Ringpfeil, M. (1981b). *Proc. V. Congress on Microbiology,* Sofia.

Ringpfeil, M. (1982a). *Energetik von Zellvermehrungsprozessen.* Vortrag 'Merseburger Technologische Tage', Merseburg.

Ringpfeil, M. (1982b). 'SCP Produktion auf der Basis von Kohlenwasserstoffen.' *Proc. Leipziger Biotechnologie Symposium,* Leipzig.

Ringpfeil, M. (1982c). 'Assessment of Industrially Performable Processes of Bioconversion of Residuals Recycled for Feed.' Suggestion to Commission of Biotechnology of IUPAC.

Ringpfeil. M. and Stephan, W. (1979). 'The Attainment of High Productivities of Cell Growth in High-Performance Fermentors.' *Proc. XXVIIth Int. IUPAC Symp.,* Helsinki.

Ringpfeil, M. and Kehr, K. (1980). 'The Aerobic Treatment of Waste Water from Livestock Production Units and the Production of Microbial Biomasses.' *Proc. II ASA Task Force Meeting,* Laxenburg (Austria).

Ringpfeil, M. and Koriath, H. (1984). 'Mikrobielle Stoffproduktion in der Landwirtschaft.' *Aus der Arbeit von Plenum und Klassen der Akad. Wiss. DDR* (in press).

Ringpfeil, M., Klappach, G. and Wünsche, L. (1978). 'Production of SCP as a Continuous Fermentation Process.' *Proc. 7th Symp. Cont. Culture of Microorganisms,* Prague.

Ringpfeil, M., Vetterlein, G., Schneider, J. and Franke, G. (1981a). 'SCP from Carbohydrates-Containing Effluents and Animal Production Wastes. *The States of the Art in the G.D.R.* Proc. Int. Symp. on SCP (APRIA), Paris.

Ringpfeil, M., Beck, D., Hadeball, W., Kreuter, Th. and Heinritz, H. J. (1981b). 'Production of SCP from Wastes in Livestock Farming.' In *Global Impacts of Applied Microbiology., Lagos 1980* (S. O. Emejuaiwe, O. Ogunbi and S. O. Sanni, eds.), Academic Press, London.

Ringpfeil, M., Hadeball, W., Scheibe, P., Kreuter, Th., Beck, D. and Karbaum, K. (1981c). 'Aerobic Treatment of Wastes in Animal Factories and Production of Proteinaceous Biomasses.' *Proc. VIth Int. Ferm Symp.,* London 1980.

Risthkov, R. S. (1982). *Vestnik Akad. Nauk U.S.S.R.* **4**, 24.

Romantschuk, H. (1975). 'The Pekilo Process: Protein from Spent Sulfite Liquor.' In *Single Cell Technology II* (S.R. Tannenbaum and D. I. C. Wang, eds.). MIT Press, Cambridge.

Sattler, K., Wünsche, L. and Prause, M. (1980). 'Möglichkeiten der Gewinnung von Koppelprodukten der mikrobiellen Eiweiss-Synthese auf der Basis von Kohlenwasserstoffen.' *Proc. II. Symp. der soz. Länder über Biotechnologie,* Leipzig.

Schneider, J. D. and Meyer, D. (1980). 'Kinetics and Stoichiometry of the Growth of C_1–Utilizing Microorganisms.' *Proc. II. Symp. der soz. Länder über Biotechnologie,* Leipzig.

Senez, J. (1983). *Acta Biotechnologica* **4**, 107.

Shaklady, C. A. (1975). 'Value of SCP for Animals.' In *Single Cell Technology II* (S. R. Tannenbaum and D. I. C. Wang, eds.). MIT Press, Cambridge.

Skulatshew, W. P. (1969). *Akkumuljaziya energij w kletke.* Izdat. Nauka, Moscva.

Smith, R. H. and Plamer, R. (1976). *J. Sci. Food Agric.* **27**, 763–770.

Steinkraus, K. H. (1980). *BioSci.* **30**, 384–386.

Stephan, W. (1980). 'Ergebnisse der Untersuchungen an Hochleistungsreaktoren.' *Proc. II. Symp. der soz. Lander über Biotechnologie,* Leipzig.

Streiblova, E. (1980). *Biol. Rundschau* **18**, 348.

Stouthamer, A. H. (1973). 'Ant. van Leeuwenhoek.' *J. Microbiol. Serol.* **39**, 545.

Taylor, I. J. and Senior, P. (1978). 'Endeavour.' *New Series* **2**, 32–34.

Viogt, B., Seidel, H., Müller, H., Beck, D., Ringpfeil, M., Riedel, M., Bauch, J., Gentsch, H. and Bohlmann, D. (1979). *Chem. Techm.* **31**, 409–411.

Washien, C. I. and Steinkraus, K. H. (1980). *BioSci,* **30**, 347.

Young, V. R. and Scrimshaw, N. S. (1975). In *Single Cell Technology II* (S. R. Tannenbaum and D. I. C. Wang, eds.). p. 564. MIT Press, Cambridge.

Institut für Biotechnologie,
Akademie der Wissenschaften der DDR,
Leipzig,
German Democratic Republic.

J. S. Chiao

Biogas Production in China

INTRODUCTION

Historical Account of Biogas Development

The development of biogas in rural China has recently been reviewed
by Liu and Chen (1981), who emphasized the compounding of feed and
choice of inoculum, and by Chen (1981), discussing the government
policies, and the organization and training of personnel for rural development. At the same time, Hsu (1981) published a textbook on biogas
technology.

The present paper aims to give a concise description of the technology of biogas fermentation, together with the recent research
development carried out in Shanghai. In addition, recent advances in the
utilization of industrial waste water and residues for biogas production,
together with some fundamental research on the methanogens, are
included in this review.

It is well known that China was the first country to utilize natural gas.
According to recent investigations, the time of this achievement was
between 100 B.C. and 100 A.D., approximately coinciding with the final
years of the Western Han Dynasty. At this time, the people in Sichuan
started to use natural gas for the manufacture of table salts from brine
(Liu, 1981). Microbial biogas production and utilization, however,
started quite recently. It was Luo Guo-rui from Sin-zhu Country in
Taiwan Province, who first started to produce biogas in the early 1920s.
He succeeded in the construction of a digester in his backyard to
produce biogas and thus provide lighting and cooking for his family. His
success led in 1929 to the establishment of the Guo-rui Gas Lighting
Co. in Swatow. Two years later he moved to Penglei Market, Xiao-si-
men in Shanghi, changing the name of his company to Zhong-Hwa-
Guo-Rui Gas Co. In ordei to propagate the advantages of biogas, he
advertised his digesters in the newspaper, offered training courses and

97

published a book on 'Practice in Zhong-Hua-Guo-Rui Gas Depot'. Statistics show that hundreds of digesters were established during this time in Shanghai, Kiangsu and other places with fairly good gas production, e.g. $0.2-0.3$ m^3/m^3—day in summer and $0.1-0.2$ m^3/m^3—day in winter. Apart from lighting and cooking, the biogas was also used in baking tea leaves, heating silk-worm incubation rooms, etc. After twenty years existence, Peinglei Market was destroyed during the China-Japan War by the Japanese, who occupied Shanghai and caused the shut-down of the biogas company. The biogas development, which had started in the 1920s thus came to an end in 1942 (Chengtu, 1982).

In the 1950s new biogas development took place mainly in the countryside, aimed at economizing and conserving fuel and organic fertilizer supplies. The favourable reaction by the farmers made expansion easy and, by March 1975, about 5 million digesters had been established and were producing biogas all over China. Of all the Provinces, Sichu   again was the frontrunner. It should be pointed out, however, that management of the digesters was a problem and that of all the digesters in the country, only $60-70\%$ were adequately managed and maintained, with the rest giving poor biogas production.

Characteristics of Chinese Biogas Production and Utilization

It was emphasized above that biogas production was limited mainly to the rural areas in China. Biogas production and utilization is therefore characterized by the following features:

(a) the majority of biogas digesters are of small-scale type, with the sole purpose of furnishing households with energy for lighting and cooking;

(b) for digester construction, local raw materials are used and for the operation, straw, green-feed, human and animal wastes serve as substrates;

(c) biogas production has to fulfil a number of purposes:
   - it has to supply energy fuel for cooking, light and feed-cooking;
   - it has to substitute organic fertilizers;
   - it has to improve the rural environment in lifting health standards through the killing effect of anaerobic fermentation on eggs of parasites and pathogenic enterobacteria

The nationwide biogas expansion during the 1950s emphasized the utilization of local materials for construction. The inevitable consequence was, however, that some of the digesters and gas holders were of poor quality, that is, not sufficiently strong and gas-tight. These 'sick' digesters gave very low gas production and were finally abandoned by their owners. It became the task of the Bioenergy Laboratory, Shanghai Scientific and Technical Association, to improve this situation by examining and rectifying the problems connected with the 'sick' digesters. They also worked out initial and weekly feed programmes in order to put the biogas production on a quantitative basis.

BIOGAS PRODUCTION AND UTILIZATION IN RURAL CHINA

Apart from simplicity in construction and economy in investment, the essential requirement for a digester is the absence of water leakage from the digester itself and gas leakages in the pipe lines, gas holder and all other connections (Science and Technology Commission, 1977).

Classification, Construction and Operation of Digesters

Digesters may be classified according to shape — spherical, global or rectangular — and according to construction material used — mortar, gravel, slab and brick (Fig. 1). The essential parts of a digester are the inlet, fermentation compartment (digester proper), gas holder, movable lid, outlet and connecting pipes. The operation of such a digester is outlined in Fig. 2.

Biogas digesters of the hydraulic type adopted in rural China have their advantages as was outlined above, but the disadvantages are also apparent. The inside pressure of the digester is inevitably high, fluctuates most of the time and consequently affects biogas production. Experiences elsewhere have shown (Cai *et al.*, 1980; Qian *et al* 1981) that digesters with separate storage tanks (Fig. 3) have a 30% higher biogas productivity compared to the hydraulic type.

Rational Compounding and Pretreatment of Raw Materials

In rural China, most families have plenty of raw materials for biogas generation, i.e. rice-straw, greenfeed, vegetable residues, human and animal excrement, garbage, etc., but in order to get plenty of gas as well

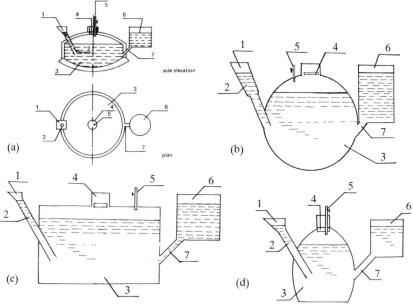

Fig. 1. Types of biogas digester. (a) Spherical, (b) global, (c) rectangular, and (d) bottle-shaped. 1 − inlet orifice, 2 − inlet, 3 − fermentation compartment, 4 − movable lid, 5 − biogas pipe, 6 − overflow compartment, 7 − connecting pipe.

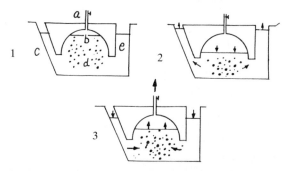

Fig. 2. Working principle of digester. a − biogas pipes, b − biogas casket, c − inlet, d − fermentation compartment, e − outlet, or overflow compartment. (1) Before biogas generation, liquor fills 2/3 of the biogas casket, liquor level is the same in the inlet, outlet and gas casket. (2) When biogas is generated, biogas collects in the casket compartment and liquid is forced into the inlet and outlet compartments. The liquid pressure in these two compartments forces the biogas towards the burners and stoves. (3) When biogas is being used, its pressure in the gas casket is reduced, the liquid comes back gradually into the fermentation and casket compartment and the liquid level fluctuates according to biogas generation and consumption.

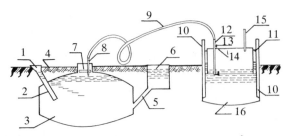

Fig. 3. Model for digester with collecting cover. 1 — inlet orifice, 2 — inlet, 3 — fermentation compartment, 4 — ground level, 5 — connecting pipe for overflow, 7 — movable lid, 8 — biogas pipe, 9 — connecting pipe, 10 — frame, 11 — guide trough, 12 — gas inlet, 13 — cock, 14 — floating cover, 15 — gas outlet, 16 — water tank.

as fast and prolonged gas production it is essential to emphasize rational mixing of the various ingredients, i.e. maintenance of the carbon and nitrogen-containing materials at the proper ratio (50–60:1) and of an adequate concentration of solids in the digester (Cu-Qiao Commune, 1977). The farmer in general adopts the following formula: human excrement 10%, animal faeces + stalk or straw 30%, and water 50%. The concentration of solids in the digester (10–20%) various in accordance with the season. It is lower in summer and higher in winter.

Pretreatment of stalk, straw and grass by composting for some time is essential for rapid biogas generation. The general practice is to compost these cellulosic raw materials (cut into short pieces), in a pit or better in the digester with alternative layers of the cut raw materials and human and animal excrement for several days in summer, and for a longer period in winter.

Filling the Digester

To start a new batch for digestion, the following procedures are adopted. First, the lid is opened, the raw materials are dumped into the digester and evenly layered. Human and animal waste is then added to about two-thirds of the digester and inoculated with the bottom layer of an old batch. The lid is put back into place and all pipe connections sealed.

Biogas starts to generate after 3–5 days in summer and 7–10 days in winter. Should the generation of biogas be too slow, the pH may be too low, and this can be adjusted using lime or vegetable ash.

Feeding and Discharging

For stable and prolonged biogas generation it is essential to add new raw materials every 10 days, after one-half to one month of operation. During each feeding, an equivalent amount of liquor is first discharged, then the raw material is added. One has to be careful during these operations to keep the liquid level above the outlet in order to avoid any escape of biogas.

Each year, the digester is discharged completely once or twice according to the need for organic fertilizers for farming. Before discharging, all materials for starting a new batch must be ready.

Daily Management

Besides regular feeding and discharging, proper daily management is also important for steady and prolonged biogas production.

(1) To speed up digestion, frequent stirring of the digester liquor is necessary in order to disrupt the crust and make the content more homogeneous. The breakdown of the crust also facilitates gas evolution.

(2) Maintenance of proper pH. When the pH of the digester liquor is below 6, a small amount of lime or plant ash has to be added. If the pH values are higher than 8, fresh animal faeces or grass and water are recommended.

(3) Control of water in digester. The water level in the digester has to be regulated because of evaporation and leakage.

(4) Inspection of leakage in all connecting pipes.

(5) Additional measures for winter. In colder districts, it is better to construct the digester 50−60 cm below the ground. Before winter sets in, a new batch has to be started with pretreated raw materials and more concentrated human and animal waste to ensure the availability of enough nutrients for the microorganisms.

RESEARCH AND DEVELOPMENT

The following experience and developmental research by the Biogas Laboratory of the Shanghai Scientific and Technical Association was

accepted for nationwide application in May 1983 (Chen 1982; Biogas Laboratory 1982).

Rectification of Constructional Defects of a 'Sick' Digester

The main defects of 'sick' digesters have been ascribed to water and gas leakage, so proper measures have been taken to correct thesse leaks: (1) thorough cleaning of the soft cracks and refilling of the latter with concrete; (2) for a leaking bottom, adding part of a second layer of concrete; (3) making connecting parts of the outlet, inlet and cover gas-tight; (4) changing the hydraulic pressure digesters into the collecting cover type (see Fig. 3), so as to lead the biogas directly into the storage tank. The inside pressure of the storage tank is kept generally at 200 mm water.

Management Plan

Considering the raw materials available in the Shanghai area and the year-round temperature, a fermentation technology and management plan has been worked out. The main features of these are as follows:

Thorough investigation and measurements have established that the daily biogas requirement for a family of 5 people is 1.2m³. With two pigs kept by the family, an additional 0.3 m³ is required. Furthermore, in the Shanghai area, pig and cow excrements and grain stalk are the chief raw materials for biogas production. Based on laboratory tests, the bio-gas capacity of these materials is about 0.3 m³ per kg of dry matter, in a retention period of 50–90 days. For example, if the daily requirement for a five-membered family is 1.5m³, then the raw material requirement for the whole year would amount to:

$$1.5 \div 0.3 \times 365 \ (\text{day}) = 1825 \ \text{kg of dry matter.}$$

Based on the daily biogas productivity within the retention period, and daily requirement, the initial feed may be calculated according to the following formula:

$$W = \frac{V}{K}$$

where W represents the inital feed (kg), V the daily biogas productivity (m³) of the digester and K the daily biogas productivity of the raw

J. S. Chiao

TABLE I
Initial feed programme

Months	1	2	3	4	5	6	7	8	9	10	11	12
Monthly ground temperature (°C)	13	12	14	15	17	18.5	22.5	25	24	22	20	15
K-value kg				0.0012	0.0015	0.0017	0.0020	0.0020	0.0015	0.0014	0.0120	0.0010
Initial feed kg				1250	1000	900	750	750	1000	1100	1250	1500
Initial solid concentration (%) 8 m³				3.1	2.5	2.3	1.9	1.9	2.5	2.8	3.1	3.8
11 m³				2.3	1.8	1.6	1.4	1.4	1.8	2.0	2.3	2.7
12 m³				2.1	1.7	1.5	1.3	1.3	1.7	1.8	2.1	2.5

Remarks:

$W = \dfrac{V}{K}$, V 1.5 m³/day,

No new batch should be made in months 1, 2 and 3.
K values and initial feeds all refer to raw materials with water content of 80%.

material in the retention period (varies with temperature and the kind of raw material).

If the digester is started by the end of April, the initial feed for a family of five people is given as:

$$W = \frac{V}{K} = \frac{1.5}{0.0012} = 1250 \text{ kg.}$$

Table I gives the initial feed for digesters with volumes of $8-12\text{m}^3$.

Based on the need for organic fertilizer, it is customary in the Shanghai area to initiate a new batch of biogas digester during the period end of April to the beginning of May. In order to maintain steady gas evolution, daily or weekly feed, determined according to the season, has to be added. The average daily or weekly feed is not the amount of total raw material minus the initial feed and divided by 365 days, but is the amount of raw material required to maintain the daily output of biogas for a digester with collecting cover. Based on experimental evidence obtained using a digester (8 m^3) with collecting cover to generate a daily output of 3 m^3 and above, it is necessary to add a daily feed of 15 kg (fresh pig waste, water content 80%) from December to March in order to meet the requirements for a family of five people (Table II). Since the biogas productivity of the raw materials is positively correlated with the temperature of the digester, the amount of feed added in summer is only half of that required in winter. In Shanghai, the temperature of a digester is generally above 20°C and weekly feeding is enough, but for digesters at temperature below 20°C feedings are necessary twice in a week to meet the required gas output. Table II gives the monthly feed for a 8 m^3 digester.

After the addition of each feed, the supernatant in the digester overflows into the storage compartment. When the temperature of the digester is above 20°C, the top crust may be broken up with a rake after removing the collecting cover. The bottom residue is easily stirred up with a disc stirrer to enable the residue to run out with the overflow. However, between the end of December and the middle of March, it is not advisable to break the crust and add river water, as the temperature of the digester would decrease abruptly. The presence of a crust in winter makes it possible to maintain a temperature of 17°C in the digester. To start a new batch for a digester with collecting cover, it is necessary to take all the crust out, but the bottom part of the liquor is kept as inoculum. It is not advisable to clean the digester completely.

TABLE II
Feed programme for an 8 m^3 digester with collecting cover.

Daily gas output (m^3)	Monthly average feed (kg), feed contains 80% water												Pigs required
	Oct.	Nov.	Dec.	Jan.	Feb.	Mar.	Apr.	May	Jun.	Jul.	Aug.	Sep.	
1	4	5	5	5	5	5	5	4	4	5	5	4	1.2
1.5	6	7.5	7.5	7.5	7.5	7.5	7.5	6	6	5	5	6	1.7
3	12	15	15	15	15	15	15	12	12	9	9	12	3.3

Remarks:
The programme is made in accordance to the climate and raw materials available in the Shanghai area.
Daily excrement of a 50 kg pig is 4 kg.

Before concluding the description of biogas technology, it is worth-while to mention the functions of biogas extension in rural China. First of all, biogas is welcomed by the farmers because it not only supplies energy, but also retains the organic fertilizer in the digester liquor. Investigations carried out in Sichuan confirm this fertilizing effect of the liquor. By comparing the nitrogen and phosphorus contents of the digester liquid from five countries and those of the ordinary manure pits after the addition of the same amount of waste and stalk, it was found that the digester liquor contains 14% excess nitrogen, 19.3% excess amino-nitrogen and 31.9% excess available phosphorous over that of the ordinary manure pits after the same period (30 days) of treatment. In addition, the digester residue is able to increase the soil humus, thus enhancing microbial activity, improving soil structure and consequently promoting higher crop yield (Inst. Soil & Fertilizer, 1978).

The extension of biogas technology in rural China also ensures better treatment of human waste. Under anaerobic fermentation conditions, the survival rate of eggs of parasites and pathogenic enterobacteria declines rapidly, demonstrating a simple way for the improvement of China's rural sanitary environment (Hook Worm Control Group, 1977; Biogas & Sc. Assessment, 1977).

BIOGAS PRODUCTION FROM INDUSTRIAL WASTE WATER

The development of biogas in the cities began in the 1960s. Industrial waste water, residues and domestic sewage are used for biogas production in cities such as Sian, Tsingtao, Fushun, etc. In the area of light industries, it was the Nan-Yang Distillery which first started with biogas production from its stillage in 1964. There are now three 200 m^3 digesters running and the daily output of biogas amounts to 10 000 m^3. In using biogas instead of coal, the production cost of ethanol is reduced by 4.43 Ren Min Bi per tonne, and the annual net profit is 170 000 Ren Min Bi. By generating biogas from the stillage, not only is energy saved, but also the BOD of waste water is reduced by 90%, and COD by 85%, at the same time yielding an organic fertilizer rich in nitrogen, phosphorus and potassium (Nan-Yang Alcohol Plant 1977). The Rong-Xian Winery in Sichuan has also succeeded in biogas production and its biogas is used for heating boilers, the generation of electricity and for household use.

Chen and Wang adopted the partial mixed anaerobic filter for the

treatment of distiller's waste, and demonstrated its high efficiency, low cost and simplicity in operation for the treatment of organic waste water of low concentration (Chen and Wang, 1980). The use of other industrial wastes for biogas production are under investigation, like the extremely high BOD waste of Shanghai Hung-Kuang Leather Factory and synthetic fatty acid wastes (Tang and Dai, 1980; Pen and Ru, 1982).

In China, many branches of light industry have rich wastewaters and consequently contribute to pollution in varying degrees. Obviously, these are rich sources of biogas production.

CONCLUSIONS

The biogas production, which started in the fifties, stressed the design and construction of digesters, considered the extension of biogas production as a matter of technology, but neglected the microbiological aspects, e.g. methanogens and non-methanogens and fermentation conditions, etc. Looking back on the past thirty years' work, it is true to say that great success has been achieved in rural China, but that an appreciable number of rural digesters (1/3) are not adequate for biogas generation. However, the recent research experience obtained in Shanghai should be able to correct such shortcomings, and will certainly place biogas generation on the basis of reliable technology. It is our belief that this experience will find wide application in the countryside.

In recent years, more attention has been paid to basic problems, and during both the 1981 and 1983 National Conferences on Biogas Microbiology, basic research has been stressed. For example, Liu *et al.* isolated 12 strains of hydrogen-producing bacteria (*Enterobacter cloacae, Escherichia coli, Serratia marcescens, Citrobacter freundii, Hafnia alvci,* and *Clostridium acetobutylicum)* from digesters and found that mixed cultures of these bacteria and enrichment cultures of methanogens, produced much greater amounts of methane with corresponding reductions in carbon dioxide (Liu *et al.*, 1980; Sun *et al.*, 1981). Zhou *et al.*, (1981) reported that the addition of *Clostridium butyricum* culture to a biogas fermentation enhanced the production of acids and methane. Feng *et al.* (1981) also stressed the functions of non-methanogens in biogas generation of pig waste. All these results demonstrated the important function of acid and hydrogen-producing bacteria on methanogenesis and are in agreement with earlier reports

(Mah *et al.*, 1977; Wolfe and Higgins, 1979). In addition, the isolation of pure cultures of methanogens such as *Methanosarcina* (Zhou and Yang, 1981; Zhou *et al.*, 1981), *Methanobrevibacter arboriphilus* and *Methanobacterium formicicum* (Quian, 1983), *Methanobacterium* spp. TC 713, TC 708 (Zhao 1983a), and *Methanococcus mazei* (Zhao, 1983b) have been successful and encourage further progress in the development of biogas production.

ACKNOWLEDGEMENTS

The author is much indebted to Professor Qian Ze-Shu, Professor Fan Qing-Shen, Professor Shen Chao-Wen, Mr. Ru Zheng-Wei and Mr. Chen Gen-Sheng for their suggestions during the preparation of this review.

REFERENCES

Biogas and Scientific Assessment of Waste Control Group, Yun-Xing Commune, Mian-Yang Country, Sichuan, 1977. *Collective Volume on Biogas,* Vol. 2, pp. 71–75.
Biogas Laboratory (1982). Shanghai Association of Science and Technology. Preliminary Study on the Biogas Technology of Collecting Cover Digester.
Cai Chang-Da, Chu, Rong-Fu, Xu Shun-Rong, Zhan Ju-Xian, Huang Shu-Nan, (1980). *Collective Volume on Biogas,* Vol. 4 pp. 11-15. Science and Technology Literature Publisher, Chungking Branch.
Chen Gen-Sheng (1982). *Studies on the Rectification of Morbid Biogas Digesters.* Biogas Laboratory, Shanghai Association of Science and Technology.
Chen, R. C. (1981). *Biomass* **1**, 39–46.
Chen Ru-Shan, Wang Zu-Xuan (1980). *Application of Partially Filled Anaerobic Column for the Treatment of Stillage Supernatant.* Kwangchow Institute of Energy Resources, Academia Sinica.
Chengtu Biogas Research Institute, Ministry of Agriculture (1982). *Advances in Scientific and Technological Research on Biogas,* No. 1, 1–4.
Cu-Qiao Commune, Chengtu, and Biogas Group, Sichuan Institute of Biology (1977). *Collective Volume on Biogas,* Vol. 2, pp. 58–60. Science and Technology Literature Publisher, Chungking Branch.
Feng Xiao-Shan, Yu Xiu-Er, Qian Ze-Shu (1981). 'Interaction Between Non-Methanogens and Methanogens in Biogas Fermentation of Pig Waste.' Paper presented at the National Conference on Biogas Microbiology.
Hook Worm Control Group, Sichuan Institute of Parasite Control (1977). *Collective Volume on Biogas,* Vol. 2, pp. 66–68.
Hsu Zeng-Fu (1981). *Biogas Technology.* Agricultural Publisher, Beijing.
Institute of Soil and Fertilizer, Academy of Sichuan Agricultural Sciences (1978). *Collective Volume on Biogas,* Vol. 3, pp. 10–14.
Liu De-Ren (1981). *Sichuan Social Science Research* **3**, 67–70; 112.

Liu Ke-Xing, Chen Guang-Qian (1981). *Global Impacts of Applied Microbiology,* Lagos, Nigeria, pp. 261—272.

Liu Ke-Xing, Xe Jie-Quan, Liao Duo-Qun, Sun Guo-Chao, Shao Ting-Jie (1980). *Acta Microbiol. Sinica* **20**, 385—389.

Mah, R.A., Ward, D. M., Daresi, L. and Glass, T. L. (1977). Ann. Rev. Microbiol. **31**, 309—341.

Nan-Yang Alcohol Plant, Honan (1977). *Collective Volume on Biogas* **2**, pp. 75—80.

Pen Wu-Hou, Ru Zheng-Wei (1982). 'Anaerobic Treatment of Synthetic Fatty Acids Waste.' Paper presented at the Sino-American Biomass-Energy Conversion Technology Symposium.

Qian Ze-Shu (1983). 'Isolation and Characterization of Methanobacteria.' Paper presented at the National Conference on Microbiology, Tianjin.

Qian Ze-Shu, Wu King-Pen, Cai Chang-Da, Zhan Ju-Xian (1981). 'Studies on Biogas Fermentation in Rural China 2. The Relationship Between Pressure and Biogas Production.' Paper presented at the National Conference on Biogas Microbiology, Cheng-Tu, Sichuan.

Science and Technology Commission, Mian-Yang District, Sichuan (1977). *Production and Utilization of Biogas.* Agricultural Publisher, Beijing.

Sun Guo-Chao, Guo Xue-Ming, Liu Ke-Xing (1981). 'Functions of the Hydrogen and Acid-Producing Bacteria.' Paper presented at the National Conference on Biogas Microbiology.

Tang King-Cai, Dai Zhi- Qing (1980). *Environmental Protection and Industrial Biogas.* (National Light Industry Environmental Protection Society, ed.). Nan-Yang District Publishing Factory.

Wolfe, R. S., and Higgins, I. J. (1979). In *Microbial Biochemistry* (J. K. Quayle, ed.). University Park Press, Baltimore, pp. 270—300.

Zhao Yi-Zhang (1983a). 'Isolation and Characterization of Methanobacterium TC 713, TC 708.' Paper presented at the National Conference of the Chinese Society for Microbiology, Tianjin.

Zhao Yi-Zhang (1983b). 'Isolation and Physiological Properties of Methanococcus mazei'. Paper presented at the National Conference of the Chinese Society for Microbiology, Tianjin.

Zhou Meng-Jin, Yang Xiu-Shang (1981). 'Enrichment Culture of Methanosarcina.' Paper presented at the National Conference on Biogas Microbiology.

Zhou Meng-Jin, Xang-Xiu, Gao-Ya (1981a). 'Effect of Inoculating Clostridium butyricum to Biogas Fermentation.' Paper presented at the National Conference on Biogas Microbiology.

Zhou Meng-Jin, Yang Xiu-Shang, Gao-Ya (1981b). 'Isolation and Identification of Pure Culture of Methanosarcina.' Paper presented at the National Conference on Biogas Microbiology.

Department of Microbiology,
Shanghai Institute of Plant Physiology,
Academia Sinica,
Shanghai,
China.

E. J. Olguín

Appropriate Biotechnological Systems in the Arid Environment

WHAT IS APPROPRIATE TECHNOLOGY?

Two different trends of thought are perceptible within the groups working in Appropiate Technology (AT) in developing countries. Some groups are still very much under the influence of the Schumacher movement and regard "AT" as being synonymous with intermediate technology on a small scale, full of simplicity and accepting in their countries the transnationals to be owners of the highly sophisticated technology.

On the other hand, there are quite a few new groups in developing countries which have been created by indigenous people without much influence from the developed countries, who consider AT as any sort of technology (i. e. high, intermediate and low technology) created locally as long as it is designed to satisfy the basic needs of the population without detriment to the environment and promotes the technological and economic independence of their countries. For example, to develop a national process for the production of antibiotics or vaccines utilizing non-polluting high technology in order to stop the drain of foregin exchange, is regarded as appropriate technology by this modern school of thought.

Thus, more emphasis on the social and political implications of AT has become apparent. The following quotations are an illustration of this concern.

Appropriate technology is neither intermediate technology nor miniaturization of production process: it is a radically new approach in which production technique becomes subordinate to social needs. (Behari, 1977)

The project [undertaking AT] should be an integrated venture wherein the organization and management, from production and collection of the raw material to its processing and distribution, is knitted together and even the growers of raw materials become the partners in production, profit and and loss. (Srivastava, 1977)

111

The main characteristic for defining appropriate technology for developing countries is therefore that it does not only imply that a choice of technology is necessary, but that it realises that changes in the existing socio-economic and political structure are essential. Only with the introduction of these changes can AT be expected to provide a means by which the regional community sustains itself.

On the other hand, the criteria which have been proposed to evaluate technology as appropriate, are based on the analysis of the problems which arise from the introduction of high technology into developing countries. The wide range of criteria suggested by authors such as Schumacher (1974), Dickson (1974), and Braun and Collingridge (1977) are very well known. However, although some of several criteria already proposed may be usefully applied in the evaluation of appropriate technology, there is still the need for an overall criterion capable of simultaneously taking into account the technical, the economic and the social aspects involved in the analysis. Thus, it seems that the approach developed by Frances Stewart (1973) concerning the appropriateness of the product linked to the method of production is a convenient overall criterion. This product-orientated approach extends the choice. Instead of taking into account only the capital-intensive vs labour-intensive and the large-scale vs small-scale criteria, this approach allows for the existing differences in capital forms, scales, skills, etc. in the different developing countries. It may be applied taking the appropriateness of the product as a starting point to which the rest of the criteria mentioned above may be applied in a flexible way to evaluate various methods aimed at the manufacture of the chosen product.

Finally, attention is drawn to the fact that AT (as commented on by Reynolds, 1978, p. 2) is not a

low-grade, primitive activity, unworthy of the attention of the highly skilled and highly educated. In fact, the development of a simple, appropriate technology often calls for a high level of basic research and the application of very sophisticated methods and equipment, in order to produce something simple, easy to operate and having adequate efficiency.

Concerning the international networking activities, Doelle (1982) has made a fundamental remark:

There is no doubt that less developed countries require help from developed countries. Researchers and advisers from developed countries should, however, realize, that requirements, resources and social structures are very different in the less developed

countries and that a proper assessment for an appropriate biotechnology for the particular country can only be carried out in that particular country and not from outside institutions.

In conclusion, scientists and technologists who are aware of the social and economic problems of underdevelopment, may form interdisciplinary teams and may promote R & D for appropriate technology at various levels. The first level could concentrate efforts on evaluating existing technology in the light of the choice of socially useful and relevant products. They can then evaluate all sorts of well known processes (i.e. "sophisticate" or "high", "intermediate" and "indigenous") to choose the more appropriate one for the area under study. The second level might be directed to the adjustment, modification and improvement of the chosen process or technology. The third level, and most important of all, is to develop locally new processes or technologies, entirely designed to be feasible and suitable for a particular situation.

GENERAL CONSIDERATIONS ON ARID ZONES

The arid zones throughout the world were practically neglected in the past until dramatic events became the focus of public awareness. The urgency of the problem was heightened by the disastrous drought of 1968—1973 which affected millions of people dependent upon the resources of arid and semi-arid lands for their livelihood. In 1977 representatives from all over the world expressed their concern at the United Nations Conference on Desertification (UNCOD). According to a study prepared in 1980 for the General Assembly, 3.3 billion hectares or 80 per cent of the world's total irrigated land, rangeland and rain-fed cropland in arid and semi-arid areas were affected by desertification. On a global level, the current annual loss of production due to desertification is in the order of $26 billion dollars (Karrar, 1982).

As a result of the current concern on desertification, various programmes have been undertaken on a global scale, such as: afforestation, water captation and conservation, rational management of plant resources, etc. However, very little has been done to apply the modern advances in the biotechnological field to improve the quality of life in the arid environment.

Biotechnology is a tremendous tool through which a wide variety of raw materials can be converted into all sort of products by means of

plenty of alternative processes. This pseudo-definition is used with the sole purpose of stressing the usefulness of biotechnology as a primary tool for effectively tackling the "food crisis", "energy crisis" or "pollution crisis".

The present article is an attempt to describe some of the biotechnological systems acceptable for the arid and semi-arid zones, not only from the technical point of view, but also from the socio-economic perspective in accordance with the general concepts of appropriate technology outlined earlier. However, we should always bear in mind that the eventual success or failure of any of the systems hereby described to improve the quality of life of people from arid zones, will not depend on their technical value, but on the socio-economic and political strategy designed for their implementation. This statement is even more valid in the case of developing countries, where ironically enough, the arid and semi-arid zones of the world are concentrated and where poverty and unequality have been endemic for a long time.

PRODUCTION OF AQUATIC BIOMASS

Production of Algal Biomass in Saline Water

Following the concept of maximum use of resources, saline water, which is an underexploited resource of arid zones at the moment, could become an essential element for biomass production in these areas. Most of the developing countries are lacking information about the vast amount of saline aquifers which underlie their desertic areas. In the case of the United States, it has been estimated, that more than 18.5 billion acre-feet have access to saline groundwater (Neary, 1981). In the case of Israel, it has been estimated that the aquifers which underlie the Negev Desert could supply 30 million m^3 of saline water annually with no depletion (Richmond and Preiss, 1979).

Thus, mass cultivation of microalgae in saline water offers a tremendous potential for production of food, feed and chemicals in the arid environment. Cultivation of macroalgae in seawater, although of great potential, is not included in the foregoing discussion.

Although various algal genera grow optimally at high salinities, there exist two genera with remarkable advantages: *Spirulina* and *Dunaliella*.

Cultivation of Spirulina. Spirulina is a well known alga, which has been

utilized in the past as a food source by the Aztecs in Mexico and which is currently grown in Lake Chad and still used as a food source by the people of Chad. Basic knowledge on this alga is extensive, and much information has been accumulated concerning its nutritional properties and the technology for its mass cultivation.

Spirulina cultivation in arid lands offers the following advantages (reviewed by Olguín, 1978):

- Productivities as high as 60—80 tonnes/ha/year may be achieved, in contrast to productivities in the range of 10—30 tonnes/ha/year in the case of conventional crops.
- Land utilization becomes more efficient when using *Spirulina* to produce protein than any conventional source. Its production require 0.03ha/tonne of protein/year in contrast to a requirement of 1.55 ha/tonne of protein/year in the case of soybeans and requirement of 452.47 ha/tonne-of protein/year in the case of beef production (range conditions), according to Leesley and collaborators (1980).
- *Spirulina* biomass shows one of the highest protein contents, ranging from 60% to 70% of the dry weight.
- It shows a high content of vitamins, three times as much thiamine and twice as much carotene as *Scenedesmus* (Becker, 1981).
- The nutritional quality as far as PER, NPU and BV are concerned are satisfactory. It is easy to digest without further treatment, since cells are surrounded by mucoproteins consisting of membrane and not cellulose as is the case in many other nutritional algae.
- There are several reports which have demonstrated that *Spirulina* is not toxic when fed to animals in certain percentages of their protein diet. Multigenerational studies in rats have also shown that the use of this algae is safe in the long-term (Chamorro, 1980).
- Concerning the technological aspects, *Spirulina,* due to its filamentous form, is easier to recuperate than the unicellular algae
- A recent report from Leesley and colaborators (1980) indicates that production of protein by means of cultivating *Spirulina,* is more energy efficient than protein production from any other conventional source. This statement applies to various energetic balances. *Spirulina* cultivation requires less energy per tonne of

produced protein, shows one of the highest energy output to energy input ratio and provides a greater amount of food energy per unit mass than any other conventional protein source.

Concerning the technology for mass cultivation of *Spirulina*, there are two levels of advance. One of these is the well established technology for its cultivation at the commerical level (2 tonnes/day) under the so-called "seminatural" cultivation carried out by the Company Sosa-Texcoco in Mexico. The second level of the technological advance has reached only the pilot-plant level and is related to the mass cultivation of this algae utilizing organic wastes as a nutrient source. It is important to emphasize that production costs are remarkably different depending on the sort of technology chosen. In the former cases, conventional technology with high energy input is involved and production costs are extremely high. These high costs have prevented the use of *Spirulina* as a supplementary protein source to counteract undernourishment in Mexico, and is only commercialized as a health food for a sophisticated market in developed countries.

However, the production of *Spirulina* is feasible at a low cost when involving non-conventional technology with a maximum of renewable energy input and utilizing organic wastes as a nutrient source. It is along these lines that mass production of algae is envisaged as an appropriate biotechnological process for developing countries.

Research for developing this low-cost technology for production of *Spirulina* is carried out mainly in Israel, India and Mexico. Optimal conditions of growth, such as a pH of 9.3, temperatures ranging from 30°C to 35°C, and light intensity from 4 klux to 5.5 klux, have been established previously (Kosaric *et al.*, 1974; Santillán, 1975; Richmond and Vonshak, 1978).

From Israel, Oron and collaborators (1979) reported on the cultivation of *Spirulina maxima* on cow-manure. They found a yield of 3g/l in terms of total suspended solids, although this yield was conditioned to a long process of degradation of the waste in the pond. Shelef (1978) reported on the advantages of using already degraded, anaerobically digested manure instead of raw manure as a source of nutrients.

Venkataraman (1980) carried out an extensive review of the work done at various research centres in India for the cultivation of *Spirulina*. Seshadri *et al.* (1980) also reported on the supplementation of anaerobically digested cattle manure (5% v/v) into the medium, in addition to

bone meal, sea-salt and sodium bicarbonate, to attain a yield of 8—10 g dry weight/m²/day.

In Mexico, our group is working on the development of a process for the cultivation, recuperation and drying of *Spirulina* at low cost, which could be appropriate for implementation at the village level where little infrastructure could be found. The process is expected to be economically feasible, providing it is within an integrated system for food, feed, fuel and chemical production, as described at the end of this chapter.

Two different alternatives have been tested. One is the cultivation of *Spirulina* utilizing raw sewage and natural saline water (Fig. 1), involving recuperation and drying. A 200 m² pilot plant has been set up jointly by IMETA (Instituto Mexicano de Tecnologías Apropiadas) and the Comisión del Lago de Texcoco. Description of the pilot plant and preliminary results have been reported previously (Olguín *et al.*, 1981; Olguín, 1982). High productivity peaks were obtained during a semi-

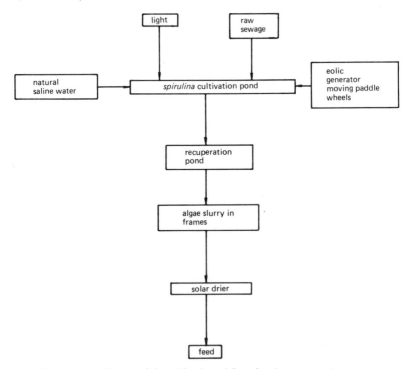

Fig. 1. Flow diagram of the cultivation of *Spirulina* in sewage at low-cost.

continuous cultivation period of 23 days at an alkalinity of 9000 mg/l $CaCO_3$, without any addition of chemicals.

On the other hand, research at IMETA has also progressed on a second alternative consisting of the cultivation of *Spirulina* on natural saline water supplemented with anaerobically digested cow manure and "Tequesquite" (natural bicarbonate scraped off the soil). In this second alternative, no recuperation is attempted, since the cultivated algae is transferred into a fish pond. Preliminary results under laboratory controlled conditions have shown that an addition of 3% (w/v) of the effluents from an anaerobic digestor processing cow manure with a retention time of 30 days is optimal (Olguín and Vigueras, 1981).

Cultivation of Dunaliella The second alga with great potential for arid zones is *Dunaliella*. The remarkable feature of this alga is that it produces intracellular glycerol in response to the osmotic stress imposed by extracellular NaCl. Chen and Chi (1981) have proposed a two-step cultivation system for this alga in seawater. Inside the first cultivation pond, rapid growth rate is promoted by controlling NaCl concentration at 1.5M. At the end of a batch-cycle, one third of the volume is transferred to a second cultivation pond, also having a 30 cm depth. The NaCl concentration in this second stage is maintained at 4.0 M in order to induce maximum glycerol production. Prescrubbed stack gas is fed continuously during the two stages.

This process utilizes a shallow-depth settler for cell recuperation followed by a solar drying bed, a screen separator and a screen press to extract the glycerol. The glycerol stream is concentrated by evaporation and finally refined by distillation to a purity of 99%.

The economic analysis made by Chen and Chi (1981) on their proposed system, has shown that it is feasible and more convenient than the petrochemical process for glycerol production. The two processes are clearly different in nature in that the algal process is capital- and labour-intensive, whereas the petrochemical process is energy-intensive. As proposed by the same authors, this process seems to be appropriate for less developed countries where labour costs are low, providing marginal land is available and climatic conditions are adequate.

Other Aquatic Plants

"Water harvesting" or the process of "collecting natural precipitation

from prepared watersheds for beneficial use" is an underexploited ancient practice used by farmers in the Negev Desert of Israel over 3000 years ago (Cooley *et al*, 1975), which requires revival and promotion in arid and semi-arid lands.

Considering that development of low-cost methods for "water harvesting" is feasible in these areas, the development of small-scale cultures of aquatic plants useful as fodder, sounds practical. There are various alternative aquatic plants worth of cultivation. Discussion hereby will focus on two neglected aquatic weeds: *Azolla* and duck-weeds (Family Lemnaceae).

Azolla lives in symbiotic association with *Anabaena azolla*, a nitrogen-fixing blue-green alga. The rate of nitrogen fixation in this association rivals that of *Rhizobium*-legume symbiosis, producing an annual nitrogen yield of 864 kg N/ha. Although primarily grown as green manure for rice, *Azolla* has also been used as fodder throughout Asia and parts of Africa. One hectare of *Azolla* can produce 540–720 kg of assimilable protein per month. It is has been fed to pigs, ducks, chickens, cattle and fish. The grass carp, *Ctenopharyng-goden idella*, and *Tilapia mossambica*, have shown preference for this aquatic weed (Lumpkin and Plucknett, 1980).

Duckweeds are tiny free-floating vascular plants belonging to three different genera, *Spirodela*, *Lemna*, and *Woffia*, which can be used as a feed supplement for aquatic and terrestrial animal stocks. The crude protein content ranges from 7 to 20% when grown in natural waters, and from 30–40% when grown in sewage-enriched waters. The yield expected varies from 10 to 12 tonnes dry weight/ha/year (Russoff *et al*, 1980).

The small-scale culturing system of duckweed seems to be a feasible alternative to providing animal protein to humans, according to the calculations given by Culley and collaborators (1981). Given a small but nutrient-rich pond of only 49 m², 40kg of plant protein per year could be produced to meet the annual feed requirements of 150 chickens, raised to feed a family of five.

Duckweeds also have a potential in waste treatment, as they have been shown to remove nitrogen at a rate of 2-4 kg/ha/day depending on weather conditions (Culley *et al*, 1981).

Finally, duckweed has been found very useful in aquaculture, since it makes a good food for grass carp (*Ctenopharyinggodon idella*) and for channel catfish (*Ictalurus punctatus*) in a diet consisting of 20% dry

duckweed in the latter case. *Tilapia* was found not to feed on duckweed (reviewed by Culley *et al.*, 1981).

BIOMASS PROCESSING

Development of biotechnological systems for the agro-industrial utilization of plants from arid zones deserves high priority. Processing of underexploited plants in marginal land useless for conventional agriculture, offers the following alternatives:

  − production of food and feed
  − production of fuel
  − extraction of valuable chemicals

Non-Conventional Crops

To produce food and feed in arid and semi-arid ecosystems has been a non-controlled ancient practice which usually has led to desertification because of overgrazing or inadequate irrigation systems. To introduce new crops does not only require advances in the agronomic field, but also the development of biotechnological processes specially designed for their integrated exploitation.

The Creosote bush (*Larrea tridentata*) has been considered an invader bush which needs controlling by herbicides or mechanical means (Herbel *et al.*, 1974). However, biotechnological developments have shown that this plant should be cultivated and exploited as source of valuable resin, with the residue going as a source of feedstuff. Although much research has been undertaken to develop an industrial process for the extraction of the resins from the leaves of the Creosote bush (Campos-López, 1979), little research has been done in the utilization of the extracted leaves as feedstuff. Adams (1970) has described a simple extraction procedure using a dilute sodium hydroxide solution and the nutritional evaluation of the extracted leaves after feeding calves for 60 days. He found a weight gain of 0.87 kg/day which compared favourable with a weight gain of 0.55 kg/day in the control batch fed with alfalfa. Olguín and collaborators (unpublished results) have tried different NaOH concentrations (from 0.5 to 2% w/v) and times to optimize the extraction procedure, together with bromatological analyses of the extracted leaves.

Table I summarizes data concerning some underexploited plants which would require only simple biotechnological processing in order to exploit their leaves or seeds as a source of protein. Research is required for the adequate ensilage of these new crops such as *Atriplex* or *Distichlis spicata* which may be badly affected by winter.

It is interesting to mention that programme for the restoration of 3000 ha of saline alkali soil left after the desiccation of Lake Texcoco and based on propagation of *Distichlis spicata,* has been a success for

TABLE I

Non-conventional crops with potential as fodder in arid and semi-arid zones

Scientific name	Common name	Protein content %/dry matter	Special characteristics	Reference
Atriplex sp.	Atriplex	12%–24% (leaves) depending on species and stage of growth	(a) Highly salt-tolerant (b) Grows well with only 150–200 mm annual rainfall (c) Resists temperatures as low as – 10°C to – 12°C (d) *A. nummularia* has been shown to be palatable. Eaten by sheep and cattle and able to sustain sheep at a ratio of 3 sheep/ha in an area of 250 mm a.r. (e) Under cultivation in saline conditions, palatability might be affected by salt being deposited on the surface of leaf.	N.A.S. 1975 Goodin, 1979
Cyamopsis tetragonoloba	Guar	24% (seeds)	(a) Fairly salt-tolerant and drought-resistant (b) Fixes nitrogen (c) Protein with a balanced aminogram which complements the amino acid deficiencies in cereals	Goodin and Northington, 1982

Table I *(continued)*

Cucurbita foetidissima	Buffalo gourd	30–35% (seeds)	(a) Seeds useful as protein supplement for monogastrics, although the bitter-tasting glycosides should be separated by washing (b) 1 ha of plants can produce 2.5 tonnes of seed (c) Seeds contain 34% of oil and provide 4.3 kcal/g	Bernis *et al.* 1978 N.A.S.
Distichlis spicata	Halophyte grass	14–19% (leaves)	(a) A remarkable grass able to grow under extreme conditions of salinity and drought (b) It is able to promote growth of cattle at a rate of 0.5 kg/day with a stock ratio of 5 head/ha with a rotation every 28 days	Llerenas and Garín 1980
Larrea tridentata	Creosotebush Gobernadora	18–20% (leaves)	(a) The leaves need to be treated with a dilute alkaline solution to eliminate surface resins (b) Calves fed with the treated leaves have gained 0.87 kg/day	Adams, 1970
Distichlis palmeris	Palmer's grass		The seeds contain 80% carbohydrates. Eaten by the Cocopa Indians in California. It can be irrigated with sea water.	Neary, 1981
Batis maritima	Saltwort		Roots and stems are eaten by the Seri Indians of Sonora, Mexico. It can be irrigated with sea water	Neary, 1981

controlling air contamination going into Mexico City. This halophyte grass contains around 14 to 19% of protein and promotes the growth of cattle to a satisfactory degree. It grows well in soils with a conductivity in the range of 40–50 mmhos/cm and with a pH in the range of 8 to 11. It is also drought-resistant (Llerenas and Garín, 1980).

Halophyte agriculture is a very promising trend for arid zones. *Atriplex* and other useful halophytes (see Table 1) may even be irrigated with seawater containing from 30 thousands to 40 thousands parts per million of salinity. Production of food crops irrigated with seawater has been reviewed by Epstein and collaborators (1979) and Somers (1978).

Improvement in Digestibility

Biotechnology can also play a key role in the more efficient use of conventional crops, especially in the arid environment where the number of useful species is limited. Processing of plants for improving their digestibility has not been fully developed. An example of the relevance of undertaking research in this field is illustrated by the case of *Prosopis cineraria*. This tree is one of the major, if not the most important, top feed for all livestock species in arid and semi-arid areas throughout the world. However, Bohra (1980) has shown that the digestibility of the leaves of this tree is very poor. While the leaves contain 14.2% crude protein, the percentage of digestible crude protein was only 3.1% for sheep and 5.5% for goats. The digestibility of the leaves containing 23.2% of cellulose was less than half that recorded for the leaves of the grass *Cenchrus ciliaris*, containing 57.4% of cellulose. The problem of low digestibility may be explained by the formation of an insoluble tannin-protein complex, which remains undigested. The development of a process for eliminating the high tannin content of the leaves could be a useful contribution to the production of livestock in arid zones.

Another interesting comment is related to the cost of the chemical treatment required for making mesquite wood (*Prosopis juliflora*) more palatable and digestible for ruminants. Currently, treatment costs are high, and research along these lines seems rewardable since chemically treated mesquite wood gives similar weight gain results and is as efficient as cotton seed hulls when supplemented up to 20% in feed rations (Goodin and Northington, 1982).

Extraction of Chemicals

Much research and experience has accumulated on the cultivation and extraction of chemicals from plants such as jojoba, guayule, and the creosote bush. On the other hand, there are many other plants from arid zones awaiting economic processes for their exploitation. Table II refers to some underexploited plants with agro-industrial protential.

The buffalo gourd (*Cucurbita foetidissima*) deserves special mention since it has potential for producing oil, starch and protein. The productivities have been estimated as follows (Bernis *et al.,* 1978). Taking into consideration that each plant produces 30 fruits per m^2 and each fruit contains about 10 g of seed, the expected yield is 3 tonnes seed/ha. Since the seed contains 30% oil, a yield of 1050 kg oil/ha is estimated. Protein yield is 450 kg/ha assuming a 15% protein content in seeds. On the other hand, starch yield is 13.5 tonnes/ha assuming 15% starch content in roots and a production of 9 kg of roots per m^2.

Extraction of chemicals from algae is also a most promising field for

TABLE II

Some underexploited plants with agro-industrial potential in arid and semi-arid areas

Scientific name	Common name	Potential use	Reference
Cucurbita foetidissima	Buffalo gourd	(a) Roots are excellent source of starch (13.5 tonne/ha) (b) Seeds are source of oil (1 tonne/ha)	Bernis, *et al.* 1978
Simmondsia chinensis	Jojoba	(a) Seeds contains 50% unsaturated liquid "wax" (esters of fatty acids and alcohols) with extremely good qualities for use in industry (lubricants, cosmetics). (b) It is a true drought-resistant desert shrub and also appears to be fairly salt-tolerant	N.A.S. 1975
Larrea tridentata	Creosotebush	Resins are valuable as antioxidants in food industry	Campos, 1979

Table II (continued)

Parthenium argentatum	Guayule	A good alternative source of rubber	N.A.S. 1975
Cyamopsis tetragonoloba	Guar	The gum extracted from the seed's endosperm has a high viscosity and five to eight times the thickening power of starch. It is used as a thickener in cosmetics and as a strengthening agent in paper products. In the petroleum industry it is used as a friction-reducing additive when drilling mud.	Goodin and Northington, 1982
Matricaria chamomilla L.	German chamomile	Air-dried flowers are steam distilled at high pressure to obtain the deep blue oil of strong odour and bitter aromatic flavour. This drug has great demand and high price in the pharmaceutical market	Singh 1969
Euphorbia eathyris	Gopher purge	Productivity of hydrocarbons in the range of 70 g/m^2/year as hexane extractables from leaves and stems	Aronson J. A. and M. Zur, 1982
Calotropis procera	Desert witch	Productivity of hydrocarbons in the range of 100 g/m^2/year as hexane extractables from leaves and stems	Aronson J. A. and M. Zur, 1982

biotechnological development in arid zones. Dubinsky and collaborators (1978) have reviewed the potential of large-scale algal culture as a source of various chemicals, mainly lipids. These chemicals are found in variable amounts depending on environmental conditions, as happens with most intracellular macromolecules. More lipids are accumulated as the algal culture becomes older or the nitrogen source approaches depletion. There are two algae which deserve special mention: *Dunaliella* spp. which is able to accumulate as much as 80% of its dry weight as glycerol, and *Botryococcus braunii* which contains 53% of its dry biomass in the form of lipids, with 19% of these lipids in the form of hydrocarbons. These latter yields compare favorably with others obtained from plants. One more interesting characteristic of both algae is that they are able to grow in both fresh water and seawater.

Dunaliella has so much potential that Williams and collaborators (1978) have proposed various hypothetical conversion routes to obtain

not only the glycerol itself and crude algal protein, but also other products derived from glycerol such as yeast biomass, mannitol, fructose, ethanol and various organic acids. The economic and technical feasibility of all these bioconversion remains to be shown for particular site conditions.

Spirulina is another algae with much potential as a source of chemicals. Richmond and collaborators (1979) have reported the following scheme for the production of protein and chemicals from this algae:

200 m² outdoor ponds = 100kg dry biomass/month

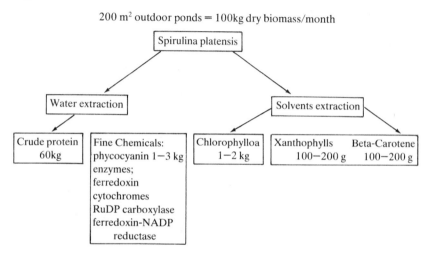

Bioenergy

Biomass as source of energy has been the concern of a large group of researchers around the world. Several reviews and a large number of reports have shown the advantages of obtaining energy from biomass, a renewable source depending on the photosynthetic capture of solar energy, and at a time when the industrial world's economy based on oil is under a severe crisis.

In general terms, two main approaches have been developed for humid and tropical ecosystems. One of these approaches is concerned with the production of "energy crops" which can be transformed either by chemical or by biochemical processes into fuels. The case of Brazil is so well known as an example of relying on biomass rich in carbohydrates to produce alcohol. The second approach is to produce biogas

through anaerobic fermentation of biomass, mainly organic waste biomass.

However, little of the accumulated information can be applied to the arid environment, where little biomass rich in fermentable carbohydrates or agricultural wastes or animal wastes is available.

Thus, research and development of appropriate biotechnological systems for bioenergy production in the arid and semi-arid zones is another challenge.

"Energy crops" have to be carefully selected for being of fast growth and attractive from the economical point of view. *Cucurbita foetidissima* (buffalo gourd) is a good candidate for research as a source for the conversion of fermentable starch into alcohol. This plant, as mentioned above, has also the attraction of producing oil and protein. The abundant foliage may be a good cellulose source for biogas production through anaerobic fermentation.

Mesquite could be a double-purpose crop: the beans and pods as source of food and feed and the wood as energy source.

Salsola kali (tumbleweed) has been selected as a useful source of energy by the University of Arizona. The National Academy of Sciences has published an excellent review of the firewood crops, including those for arid lands (National Academy of Sciences, 1980).

On the other hand, biogas production has to be designed also under different constrains: animals are usually not in feedlots and another source of biomass, different from animal waste, has to be the substrate (phytomass). Temperatures go to extremes during winter and summer. Fresh water is scarce and saline water may be the only type of water available.

Hundreds of pages have been produced describing technical aspects for building and operating anaerobic digestors at small scale in developing countries. These have been directed mainly towards the bioconversion of animal waste into fuel and fertilizers. However, very few reports have been produced evaluating the technical and socio-economical impact of the promotion of these systems in rural societies of the countries. Many important questions require a precise answer

(a) Would large cooperative units rather than small unifamiliar units be a better option capable of giving opportunities to the poorest peasants and also capable of operating more economically and contributing in a larger extent to the energy deficit?

(b) Would continuous flow digestors be more convenient than batch digestors, regardless of the human factor concerning daily attention? The Latin American Organization for Energy (OLADE) launched a large programme for the construction of 100 biodigesters in various countries in 1981. The evaluation of this programme has shown among other things that the batch design has the highest success (63%), apparently due to the fact that the human factor was less involved (Monteverde, 1983).

For the arid environment in developing countries, where extensive animal raising predominates, feed production has to be solved before feedlot practice sounds feasible, and the use of animal waste will turn practical. However, the cultivation of specific crops for anaerobic digestion would have the great advantage of providing fertilizers besides fuel, in contrast to the current practice of burning directly the local phytomass. If this could be done, a basic principle, concerning the usefulness of integrated biogas farming systems to reduce pressure on natural ocurring fire-wood resources, and to reduce the cost of buying inorganic fertilizer, would be accomplished as pointed out by DaSilva and collaborators (1980).

INTEGRATED SYSTEMS

In the preceeding sections, biotechnological systems for the production or processing of a particular microorganism or plant from arid lands, have been described. Although each has a value in itself, the economy of the processes improve significantly when they become integrated within larger systems, where end-products of a single system serve as a raw material of a subsequent one. Furthermore, the recycling of energy and nutrients in these integrated systems allow large productivities and diversification of products which are not usual or possible otherwise in the arid environment.

Thus, integrated systems should be preferred providing the following requirements could be satisfied:

(a) The social organization of the community is mature enough to run the integrated system properly.
(b) There is sufficient capital for initial investment. This factor calls for the necessity of cooperatives undertaking the management, rather than individuals.

(c) There is an adequate level of expertise on the permanent technical staff. For rural communities in developing countries, intensive training-programs are essential, once a demonstrative unit has been set up.

Several integrated systems either at small- or at a large-scale have been described previously (National Academy of Sciences, 1981). However, all of them have been designed for tropical or humid ecosystems, and most of them, which are currently in operation, are located in Asia. Thus, if to increase productivity in arid lands is a difficult task, full implementation of integrated systems in these areas is a real challenge.

Two different integrated systems for arid zones are described, but it should be realized that specific systems should be designed to fit the needs and the available resources in each particular situation.

Fig. 2 outlines an integrated system for the production of food, feed, fuel and chemicals in arid zones that has been promoted by the Mexican Institute for Appropriate Technology (IMETA) in Mexico. The system takes advantages of most available resources in arid lands: solar energy, saline water, harvested rainwater, organic waste, non-conventional crops, halophyte plants or microorganisms, and energy crops.

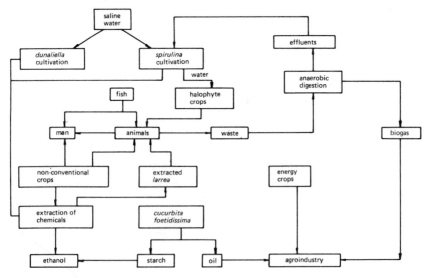

Fig. 2. Integrated system for the production of food, feed, fuel and chemicals in arid zones.

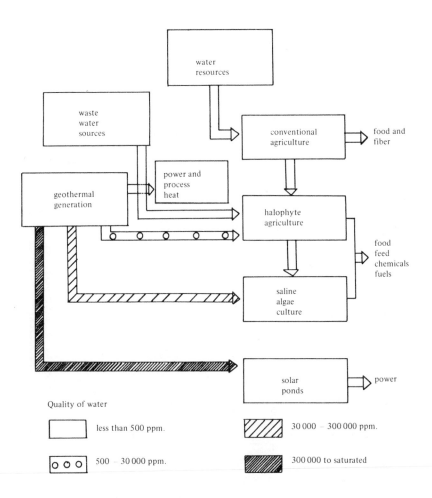

Fig. 3. Proposed cascading scheme for water reuse in arid lands. (After Kingsolver, 1982)

The result of such an integrated system can lead to a wide diversity of products:

- food: fish, cattle, pigs, chickens
- feed: *Spirulina* as source of protein, halophyte crops, extracted *Larrea,* buffalo gourd seeds
- fuel: biogas and ethanol
- chemicals: pigments from *Spirulina,* glycerol from *Dunaliella,* anti-oxidant from *Larrea.*

Although there is a fair complexity in the system, maximum use of renewable energy is involved to make it economically feasible such as in the already described low-cost process for the cultivation of *Spirulina* involving agitation by eolic energy and drying by solar dryers. Since the system is labour-intensive, the generation of employment may be another important benefit.

So far, IMETA has reported on the development of few of the modules which form the whole integrated system (see previous sections). However, the expected technical and economical feasibility of the entire system can only be shown in the long term, providing adequate financing is available.

Finally, an integrated system pioneered by the Bioenergy Research Facility of the University of Arizona is presented (in Figure 3). This system is based on an integrated water reuse cascade to produce food, feed, fuel and also chemicals (Kingsolver, 1982).

The level of complexity of this system, as far as available technology is concerned, is high. Long term, high quality and heavily financed research is required before the full implementation of the system is possible. However, long-term planning in developing countries should take into serious consideration public investment along these lines.

REFERENCES

Aronson, J. A. and M. Zur. (1982). 'Bioenergy Research in Israel: Current Assesment.' *Arid Lands Newsletter* **16**, 11–14.
Behari, B. (1977). In 'Technology for the Masses.' *Invention and Intelligence* (Jan.-Feb.).
Becker, E. W. (1981). 'Algae Mass Cultivation Production and Utilization.' *Process Biochem.* Aug/Sept.
Bohra, H. C. (1980). 'Nutrient Utilization of *Prosopis cineraria* (Khejri) Leaves by Desert Sheep and Goats. *Ann. Arid Zones* **19** (112), 73–81.
Braun, E. and D. Collingridge (1977). *Technology and Survival.* SISCON, Butterworths.

Campos-Lopez, E., T. J. Mabry and S. F. Tavizon (1979). *Larrea. CIQA* Mexico.

Chamorro, G. (1980). 'Multigeneration Studies of *Spirulina* in Rats. *Third International Conference on the Production and Use of Microalgae.* Trujillo, Perú.

Chen, B. J. and C. H. Chi (1981). 'Process Development and Evaluation for Algal Glycerol Production.' *Biotec. Bioeng.* **23** (6), 1257.

Cooley, K. R., A. R. Dearick and G. W. Frasier. 1975. 'Water Harvesting: State of the Art.' In *Watershed Management Symposium.* A.S.C.E., U.S.A.

Culley, D. D., E. Rejmankova and J. B. Frye (1981). *Production, Chemical Quality and Use of Duckweeds (Lemnaneae).* In *Aquaculture Waste Management and Animal Feeds.* World Mariculture Society.

DaSilva, E. J., W. Shearer and B. Chatel (1980). 'Renewable Bio-Solar and Microbial Systems in "Eco-Rural" Development.' *Impact of Science on Society* **30** (3), 225.

Dickson, D. (1974). *Alternative Technology and the Politics of Technical Change.* Fontana, Collins.

Doelle, H. (1982). 'Appropriate Biotechnology in Less Developed Countries.' *Cons. Recyc.* **5** (1), 75.

Dubinsky, W., T. Berner and S. Saronson (1978). 'Potential of Large-Scale Algal Culture for Biomass and Lipid Production in Arid Lands.' *Biotech. Bioeng. Symp. No. 8,* 51.

Epstein, E., R. W. Kingsbury, J. D. Norlyn and D. W. Rush (1979). 'Production of Food Crops and Other Biomass by Seawater Culture. In *The Biosaline Concept.* Hollaender, A. (Ed.). Plenum. Publishing Co.

Goodin, J. R. (1979). 'Atriplex as a Forage Crop for Arid Lands.' In *New Agricultural Crops.* Ritchie, G. A. (Ed.). AAAS, Colorado.

Goodin, J. R. and D. K. Northington (1982). 'Guar Shows Potential as Arid Lands Crops.' *Arid Land Plant Resources* **4** (1), 1.

Herbel, D. H., R. Steger and W. L. Gould (1974). *Managing Semidesert Ranges of the Southwest.* New Mexico State University. Cooperative Extension Service. Circular 456.

Karrar, G. (1982). 'Editorial.' *Desertification Control* (6) April.

Kingsolver, B. (1982). 'The University of Arizona's Bioenergy Research Facility.' *Arid Lands Newsletter* **16**, 15.

Kosaric, N., H. T. Nguyen and M. A. Bergougnou (1974). 'Growth of *Spirulina maxima* Algae in Efluents from Secondary Waste Water Treatment Plants.' *Biotech. Bioeng.* **16**, 881.

Leesley, M. E., T. M. Newsom and J. D. Burleson (1981). 'A Low Energy Method of Manufacturing High-Grade Protein Using Blue-Green Algae of the Genus *Spirulina.'* ASAE *National Energy Symposium.* Vol.3 *Agricultural Energy*, 619—623.

Llerenas, A. and M. Garin. (1980). 'Una alternative para integrar a la productividad algunas de las areas altamente salino-sodicas del pais.' *Second Interamerical Conference on Technology of Salinity and Management of Water.* Cd. Juárez, Chihuahua, México. December 1980.

Lumpkin, T. A. and D. L. Plucknett (1980). *Azolla:* Botany, Physiology and Use as a Green Manure.' *Economic Botany* **34** (2), 111.

Monteverde, F. (1983). Personal Communication.

National Academy of Sciences (1980). *Firewood Crops.* National Academy Press. Washington.

National Academy of Sciences (1981). *Food, Fuel and Fertilizer from Organic Waste.* National Academy Press, Washington.

Neary, J. (1981). 'Pickleweed, Palmer's grass and Saltwort.' *Science* **2** (5), 38.

Olguín, E. J. (1978). 'Appropriate Technology: The Case of Single Cell Protein (SCP) and Biological Upgrading of Wastes.' *Research Fellow Report.* Technology Policy Unit. University of Aston in Birmingham, U.K.

Olguín, E. and J. M. Vigueras (1981). Unconventional Food Production at the Village Level in a Desert Area of Mexico. *Second World Congress of Chemical Engineering.* Montreal-Canada (5—8 October).

Olguín, E., A. Becerra, R. Leyva and J. M. Vigueras (1981). 'A Low Cost Process for the Cultivation of *Spirulina* in Sewage at the Rural Level.' *World Conference on Aquaculture Venice-Italy* (20—25 September).

Olguín, E. (1982). 'Conversion of Animal Waste into Algae Protein Within an Integrated Agriculture System.' *Proccedings of the Seminar Microbiological Conversion of Raw Materials and By-Products of Agriculture into Protein, Alcohol and Other Products,'* Novi Sad, Yugoslavia.

Oron, G., G. Shelef and A. Levi (1979). 'Growth of *Spirulina maxima* on Cow-Manure Wastes.' *Biotech. Bioeng.* **21** (12), 2169.

Reynolds, G. F. 1978. 'Technologies Appropriate in Rural Development.' *Notes Provided at the Seminar on Technology in Rural Development.* University of Readings, 3—6th April 1978.

Richmond, A. and A. Vonshak, 1978. 'Algae — An Alternative Source of Protein and Biomass for Arid Zones.' *Arid Lands Newsletter* (9). **1.**

Richmond, A. (1978). 'Optimization Studies of Algal Biomass Production on Brackish Water for Industrial Purpose.' *Annual Report.* The Joint German Israeli Research Projects.

Rusoff, L. L., E. W. Blakeney and D. D. Culley (1980). 'Duckweeds (Lemnaceae Family): A Potential Source of Protein and Amino Acids.' *J. Agric. Food Chem.* **28**, 848.

Santillán, C. (1978). 'Avance en la Sistematización del Cultivo Seminatural de *Spirulina. 11 Coloquio Franco-Mexicano de Alga Spirulina.* Mexico, D. F.

Schumacher, E. F. (1974). *Small is Beautiful.* Abacus edition by Sphere Books Ltd.

Shelef, G. (1978). 'The Combination of Algal and Anaerobic Waste Treatment in a Bioregenerative Farm System. *Proceeding of International Workshop on Bioconversion of Organic Residues for Rural Communities.* Guatemala, C. A.

Seshadri, C. V. and S. Thomas (1980). 'Mass Culture of *Spirulina fusiformis* Using Low Cost Methods,' *National Workshop on Algal Systems, Proceedings.* October 3/4. Madras, India.

Somers, G. F. (1978). 'Production of Food Plants in Areas Supplied with Highly Saline Water: Problems and Prospects.' In *Stress Physiology in Crop Plants.* Musell, H. and R. C. Staples. (Eds.) Wiley Interscience, New York.

Srivastava, J. C. (1977). In *Technology for the masses.* Invention and Intelligence (Jan-Feb).

Stewart, F. 1973. 'Choice of Technique in Developing Countries.' In *Science Technology and Development.* Cooper, C. (Ed). Frank Cass. London.

Venkataraman, I. V. (1980). 'Algae as Food/Feed. A Critical Appraisal Based on Indian

Experience.' *National Workshop on Algal Systems Proceedings.* October 3/4. Madras, India.

Williams, L. A., E. L. Foo, A. S. Foo, I. Kuhn and C. G. Hedén (1978). 'Solar Bioconversion Systems Based on Algal Glycerol Production.' *Biotech. Bioeng. Symp.* **8**, 115.

Instituto Mexicano de Tecnologias Apropiadas,
Apartado Postal 63—254,
O200 México, D. F.,
México.

G. A. Zavarzin

Microbiology: Global Aspects

Global microbiology deals with the study of the total activity of microbial communities in their natural habitats and their role in the biosphere. The need to develop this interdisciplinary branch of science stems from the significant disturbances occuring in the biochemical cycles, whose principal stages are dependent on microflora activity. Unlike laboratory investigations, the subject of global microbiology is the superspecies microbial community as it affects the environment. The need to develop methodical approaches for estimating the total activity of microorganisms within the natural formation is discussed here, with the study of gas exchange, especially the formation of the soil-air balance, as a suggested promising approach.

GLOBAL CYCLES

The role of microorganisms in maintaining conditions on earth was generally realized during the blossoming of bacteriology towards the end of the 19th Century when, in the course of a relatively short period of time, the principal physiological groups of microorganisms were described. The subsequent successes of biochemistry and molecular biology, using microorganisms as their basic model, being fundamentally novel, have tended to obscure this traditional field. A pronounced renewal of interest in the role of microorganisms in the biosphere has stemmed from conclusions drawn in a number of related disciplines dealing with the earth sciences and representing, in fact, an interdisciplinary branch of learning. This interest was due to the supposition that the development of human activity is capable of so strongly influencing the development of the biosphere in the not so-remote future as to radically alter the conditions of life on earth. Establishing a balance according to the leading cycles of elements, first of all biogenic, proved a convenient form of evaluating these influences. Owing to its teaching on the role of microorganisms in the cycling of elements, general microbiology proved itself quite prepared for such an approach.

The important thing at the present moment is that the scale of anthropogenic changes has become comparable with that of natural processes, and for certain biogenic elements may become equal to it within the next few decades. In these circumstances the microflora, being the most labile part of the living world, may compensate most rapidly for the unfavourable changes caused by man's activity, especially owing to the ability of microorganisms to affect chemical transformation.

Changes occurring in the elemental cycles affect first of all the atmosphere and, through it, the climate.

Hence, the obvious urgency of the tasks of global microbiology which by its very nature should yield material for generalization.

The overall purpose of available forecasts of man's economic activity is the establishment of the limits to the application of existing technologies and tendencies. They are based on evolutionary processes but not revolutionary changes; the latter are essential and inevitable, but hardly predictable, although one can pinpoint areas in which they are probable.

A global forecast of change in the biosphere (Moiseyev *et al.*, 1979) reveals its general developmental tendencies. Though one may question quantitative evaluations, the general character of the changes concerned are revealed by different approaches. The essence of these changes is as follows. The impending increasing consumption of energy is based, of necessity, on extractable carbon fuels, whose mass is comparable with the mass of oxygen in the atmosphere. The binding of oxygen into carbon dioxide determines the decrease of its content from the nominal 21 per cent to 16.5 per cent by volume. The increase of carbon dioxide is described by a complex which is the result of the compensating effect of photosynthesis and the buffer system of the ocean. Changes in the carbon dioxide content in the atmosphere is a well established fact. An increase of carbon dioxide content should be followed by an increase of temperature which, however, has not so far occurred. Apparently the carbonate system of the ocean lags behind the production of carbon dioxide.

Increased photosynthesis is a compensating process favoured by an increasing concentration of CO_2 in the foliated zone. It should be borne in mind, however, that an increase in the global CO_2 concentration is not equivalent to that of the carbon dioxide in the photosynthesizing zone, the latter being affected by soil respiration determined by the

activity of microorganisms. The increasing use of fertilizers, nitrogen fertilizers in particular, whose scale of manufacture is approaching that of natural nitrogen fixation, tends to increase photosynthesis. This is the reason why the forecast points to the doubling of the vegetation cover's density. On the other hand, phosphorus drops to a minimum as it is scattered because of its poor assimilation by plants (25%), and is immobilized in anaerobic deposits on the ocean floor. It is thus phosphorus which may restrict photosynthesis, and then any increase of available nitrogen will result in the enhanced activity of microorganisms influencing gaseous nitrogen compounds. Local changes in the concentration of nitrogen oxides are capable, along with carbon dioxide, of upsetting the temperature regime and consequently the circulation and redistribution of moisture. Forecasts of such changes have been given by M. I. Budyko (1980).

All these thoughts point to the validity of global microbiology: changes in the intensity of microbial processes may indeed cause a restructuring of the planet's entire economy to suit the new conditions.

The newly emerging problems may be designated according to the main gas-forming component: (a) the problem of CO_2 (SCOPE—13, 1979; Bach *et al.*, 1980); (b) the problem of ozone, which predominantly concerns photochemistry (Talrose *et al.*, 1979); (c) the problem of N_2O; and (d) the problem of sulphur compounds (SCOPE—7, 1975; Hahn, Junge, 1977; Rosswall, 1978). In the absence of data essential for a valid forecast, each of them sounds a warning bell since they may substantially influence the biosphere. The rest are associated with the carbon cycle as the leading cycle.

Thus, emerging as the priority requirement is a more profound and clear understanding of the role of microorganisms in the carbon cycle, since the role of other creatures in the destructive part of the cycle is negligibly small. The microorganisms' role is comprised of two components: action and inaction, with the latter component probably being the more important.

The biological productivity of the ocean corresponds to one-quarter that of the land, and the principal reservoir of organic matter in the ocean is scattered soluble organic matter which, owing to its low concentration, is unaffected by bacteria.

In the aspect under consideration we are less interested in the concentration of CO_2 in the atmosphere, which is determined to a large extent by the physico-chemical equilibrium of the troposphere and the

ocean and the reservoir of inorganic carbon in general. The organic forms of carbon important for the present gas balance are only those whose age is less than 10 years. Thus, the tremendous reservoirs of scattered organic matter of mountain rocks (6 600 000 Gt of carbon) and dissolved organic matter of the deeper portions of the ocean (1 620 Gt) may be disregarded. Taking into account the fact that the organic matter of the surface layer of the ocean comes to only 30 Gt, we must turn out main attention to land-based ecosystems.

Here one may single out the reservoirs of rapidly exchangeable (up to 10 years) and slowly exchangeable (up to 1000 years) organic matter (see Table I). Since the oxidising decomposition of organic matter is estimated by calculating, rather than obtaining experimentally based differences, no data are available. There are also other analoguous calculations with other assessments of the boundaries and capacities of the reservoirs, but their correlation turns out to be nearly equal. Our task here is to assess the comparative significance of experimental studies. The traditional notion according to which the chief producer of oxygen on earth are tropical forests, has been found to be not quite true. The intensity of tropical photosynthesis is, of course, much higher than in the temperature or the boreal belt, but the speed of organic decomposition of the soil is also much higher. The total content of organic matter both in the biomass and in the soil seems to be important for evaluating oxygen production. The reservoir of dead organic matter in the soil turns out to be nearly three times greater than in plants and it is better preserved under lower temperature. As a result, the taiga makes the same contribution to the oxygen-carbon dioxide balance as do tropical forests. Hence the far from trivial conclusion that for

TABLE I

Organic carbon in different ecosystems, in Gt C (after Chan, Olson, 1980)

Ecosystem	Exchangeable reservoir		Exchangeable flow	
	Rapid	*Slow*	*Rapid*	*Slow*
Forest north of 30°N	70	530	10	8
Forest south of 30°N	53	580	11	9
Non-forest systems	37	490	10	8
Surface layer of ocean	30		23	

preserving the conditions for life on earth the forests of the temperature climate are no less important than tropical ones, due to the lower activity of the microflora. The removal of the biomass and the substitution of technical processes for microbiological ones, for example, under the "green energy" programme, is admissible for the tropics alone (on the condition of the return of mineral nutritive elements) and is doubtful for countries with temperature climates.

The picture is still more striking when assessing hydromorphous formations. Their carbon content has not been established with any degree of certainty which the upper and lower values differ by several orders of magnitude. According to the highest estimate they account for up to 40 per cent of organic carbon. In marshlands microbial processes are blocked. Thus, "young carbon", whose age corresponds to that of atmospheric oxygen, originates from the vegetation biomass which microorganisms were unable to decompose. It can be fittingly noted here that mineral fuels corresponding to oxygen long bound in geochemical processes were also formed under hydromorphous conditions.

The importance of hydromorphous formation stems also from the conceptual model of microbial processes. As is known, the respiration of the soil decreases by several orders when it is moistened, with a corresponding drop in the expenditure of oxygen. As a result of its poor solubility, the soil's organic content available to aerobes already being comparatively low, and oxygen consumption exceeding its supply from outside, conditions become anaerobic and the anaerobic microflora begin an elective readaptation to the new conditions. This involves the evolution of such products of incomplete reduction as, for example, nitrous oxide. Such transitional processes are manifested much more intensively in the soil than in bodies of water.

As a result of the activity of anaerobic bacteria the carbon dioxide is partially returned without any expenditure of oxygen. The so-called primary anaerobes produce organic acids, carbon dioxide and hydrogen. The former two components render a leaching effect on bedrock minerals and it is from this that there emerges a link between the carbon cycle and the geochemical conversions of metal. Secondary anaerobes utilize first of all the products of the primary anaerobes — hydrogen and fatty acids. The secondary anaerobes utilize oxidized compounds as an electron sink. In addition to the low specific processes of the reduction of nitrates, iron oxide and manganese, a determining role in the cycle of elements on earth is played by the reduction of sulphur, nitric oxides

and carbon dioxide. Secondary anaerobes are the basic producers of the atmosphere's minor gaseous components and these gases may get into the atmosphere when the anaerobic zone is close to the surface, which happens during bog formation.

So let us draw conclusions from this section and establish priorities.

The subject of global microbiology exists and is of vital importance to mankind. Its most urgent preoccupation is the effect of micro-organisms on elements with an aerial form of migration, especially gases, since, because of the relatively small reservoirs and mixing in the atmosphere, the effect of microorganisms assumes a global character most rapidly.

Rather than the unique, it is the common, most massive processes that are of prior importance for global microbiology. Since the carbon cycle is the leading one in the biosphere, the significance of organo-trophic microorganisms for global microbiology seems to be most important.

The formation by microorganisms of the gas in the soil which undergoes exchange with the atmosphere, is the direct means of their influence on gaseous components. Hence, the gas exchange of the soil appears to be the most important sphere of interest of global microbiology.

The contribution of microorganisms to the formation of the weathering crust, the diagenesis of mineral in sediments, traditionally studied by geological microbiology, are of doubtful global importance.

INTERACTION OF CYCLES

The carbon cycle undoubtedly serves as the motive mechanism in geochemical transformations on the earth's surface. Photosynthesis plays a primary role in this cycle, and bound to it the oxygen cycle. In the history of the earth the oxygen cycle had been connected not only with the carbon cycle, by returning carbon dioxide, though this constitutes the principal path of closing the cycle. Excessive oxygen of photosynthesis was included in nearly equal shares in the fate of iron and sulphur. Almost 50 per cent of this oxygen was used in the formation of the iron oxide deposits which, as is known today, are formed in considerable amounts by iron-depositing bacteria. But the formation of iron oxide deposits takes place, generally speaking, via the noncyclic path, uncharacteristic of biological processes. Iron has no

aerial form of migration. Although the biogenic reduction of iron oxides by microorganisms has been indeed established, the significance of this process so far seems to be small, and throughout the history of the earth iron oxide has accumulated. This is a very vague concept, since the reduction of iron during anaerobic biological conversions of organic substances does take place, but does not occur in such a highly specialized form as in the cycles of elements with the serial form of migration.

The next of the important reservoirs of bound oxygen is the sulphates of the sea. They correspond to 40 per cent of the excessive oxygen of the Phanerozoic. The motive mechanism of the sulphur cycle is bacterial sulphate reduction at the expenses of organic matter, which is responsible for 90 per cent of the sulphide-sulphur formed on earth. The sulphur cycle is closely linked with the behaviour of iron, since hydrogen sulphide reacts with iron, forming iron sulphides, and becoming involved in the cycle of geological transformations. The speed of this process is determined by the accessibility of sulphate, the presence of iron, and the exchange velocity of interstitial water in places of bacterial growth. The formation of pyrite serves as a blind offshoot of the cycle's reduction branch. Under aerobic conditions excessive hydrogen sulphide is oxidized by thionic or anaerobic phototrophic organisms in the light, while sulphates of the sea serve as the blind offshoot sink for the oxidizing branch. Only a very small part of the hydrogen sulphide may get into the atmosphere, where apparently methyl sulphides are of greater importance, especially dimethyl sulphide, the flow of which into the atmosphere, just as of hydrogen sulphide, is estimated at 10 Mt a year, that of carbonyl disulphide at 5 Mt a year and that of carbonyl sulphide at 2.5 Mt a year. The exact path of formation of these compounds in the biosphere is not clear, though it is known that dimethyl disulphide can be formed, for example, by pseudomonads. In the atmosphere these reduced compounds serve as a source for the formation of sulphates and sulphur dioxide, which return to earth and take part in the formation of Young's Layer.

The sulphur cycle has been studied in detail for a long time in the U.S.S.R. by M.V. Ivanov and results of these studies are available. One should recognize as the basic conclusion of recent years the doubling of the sulphur run-off from the continents during the last half century, with all the consequences of leaching and decalcification. The further development of coal power generation is bound to result in a 6 to

10-fold increase of sulphur run-off. The sulphur cycle thus illustrates well the scale of anthropogenic influence on the biosphere.

An interesting discovery of recent years has been the hydrogen cycle (Kondratyeva, Gogotov, 1981). Hydrogen is produced in diverse anaerobic processes of organic matter decomposition by primary anaerobes and as a side product of photosynthesis, particularly in blue-green algae. The evolution of hydrogen is a means of electron sinking off the reducer, with protons serving as the acceptor. An H_2 reservoir of any significance is lacking. Well pronounced in the cycle is the anaerobic oxidizing branch. Hydrogen may be oxidized during anaerobic photosynthesis both by specialized groups of bacteria and by many algae. Hydrogen proves a universal reducer for secondary anaerobes, reducing compounds of nitrogen, sulphur and carbon dioxide. At the same time the group of hydrogen bacteria oxidizing molecular hydrogen with oxygen turns out to be exceptionally varied. The hydrogen cycle quite possibly contains a key to the understanding of the interaction of microbial systems.

And, finally, the nitrogen cycle in nature. Constantly in the centre of recent attention, the nitrogen cycle has been given a number of new interpretations. Firstly, the number of known symbiotic and free living nitrogen-fixing bacteria is growing rapidly, for the genetic weight of this property is comparatively small and if the bacteria are prepared to pay the price in energy for nitrogen fixation, including protection against oxygen, this capacity may be manifested in different physiological groups. Secondly, field methods for determining the intensity of nitrogen fixation have provided an immediate idea of the scale of the natural process involved in global microbiology. Both these factors point to the greater scale of bacterial nitrogen fixation than formerly assumed. The biological fixation of nitrogen is capable of ensuring a balanced development of the ecosystem.

Furthermore, the attitude to oxidized forms of nitrogen has changed fundamentally. The supposed increase of nitrous oxide emission in connection with the use of fertilizers and the discovery of the aerobic path of its formation during nitrification caused an avalanche of investigations. Nitrous oxide is formed by bacteria only, and the photochemical run-off drastically upsets the equilibrium of gas occuring in the atmosphere in the concentration of several parts per million ("p.p.m. gases"). R. I. Pyodorova's discovery of the inhibition of nitrous oxide conversions by acetylene provided a convenient field method for

studying the exchange of this compound. It is now important to find the ways, and determine the scope, of bacterial removal of nitrous oxide.

Classical denitrification with the formation of H_2 is no longer the bugbear of the nitrogen cycle, but has turned into a means of attaining an ecologically safe state when bound nitrogen compounds enter from outside.

Anthropogenic effects cause a doubling of the flows in the nitrogen and sulphur cycles and a substantial change in the carbon cycle on account of the combustion of mineral fuels and microbiological changes in the organic content of the soil during its economic utilization.

What, then, are the basic interaction mechanisms of the cycles? When looking at the interaction of cycles one should single out several blocks in regard to photosynthesis − the leading process on the earth's surface.

Photochemical reactions combine into a complex system the gaseous components of predominantly biogenic and, with the exception of O_2 and CO_2, bacterial origin. The introduction into the circulation of hydrocarbon halides (freons) inaccessible to microorganisms, immediately presupposes the disruption of photochemical equilibrium and the absence of compensating biological reactions.

The photosynthetic block not only determines the oxygen-carbon dioxide balance, but also the absorption of elements in a correlation needed for biomass build-up. This block of assimilation reactions is equalized by predominantly aerobic microbial destruction carried out predominantly by mycelial organisms. During aerobic destruction some gases (for example, methyl sulphides) escape into the atmosphere, while difficult to decompose organic sustances (humus) go into the pedosphere. The block comprising photosynthesis and aerobic destruction is approximately balanced, as this follows from balance calculations and the correlation between the subsidence and respiration of the soil. The closed state of this block indicates that the total effect may be small, but due to the fact that the flows in this block are particularly great, even a slight inbalance could lead to striking consequences.

The block of catabolic anaerobic decomposition depends on the activity of bacteria forming clear-cut pairs of secondary anaerobes and lithotrophs. This block is responsible for most of the gases entering the atmospheric cycles (N_2O, NO_2, H_2S, CH_4 . . .). Through the sulphur cycle it is most closely bound to the formation of minerals.

The block of geological reactions is determined by the formation of

minerals and physico-chemical balances in water. It is here that the cycle of biogenic elements links up with the fate of metals. Alkaline-earth metals become involved with the carbonate system, while the formation of biogenic sulphides turns out to be fundamentally important for the majority of other systems.

An examination of the interacting blocks clarifies the catalytic role of lithotrophic microorganisms, which is far out of proportion to their number in nature.

The interaction of cycles makes up an interdisciplinary field of knowledge such that the assessment of the contribution of microbial processes within a difinite natural formation is an absolutely necessary task of global microbiology.

MICROBIAL SYSTEMS

An examination of the interactions between cycles demonstrates that different physiological groups of microorganisms are closely interrelated through exchange of vital products and represent an integrated supra-species system. This system of necessity includes organisms with different functions and in essence cannot be homogeneous. The system's stability depends on the possibility of replacing one or another component by an organism of similar basic function but different in adaptation. For example, even such specialized groups as nitrifying, methane-oxidizing and methane-forming bacteria are represented by a certain number of forms, similar in catabolism, but rather different in other aspects. To the bacteriologist this diversity seems excessive and uneconomical.

The most important means of uniting bacteria in a supra-species system are trophic interactions, and the formation of pools of extra-cellular components has turned out to be a special feature of trophic interactions in bacteria. This refers both to pools of vitamins and other growth factors, and also pools of exoenzymes-hydrolases and low molecular substrates. As a result, the microorganisms' kinetic characteristics acquire special importance (Pechurkin, 1978). Appearing here within the microbial system is a certain possibility of exchanging genetic material which results in the filling of free niches, and also of phenotypic changes in the system (Slater, Goodwin, 1980; Konings, Veldkamp, 1980).

Even a more or less homogeneous group of bacteria possessing

aerobic organotrophic metabolism, whose main product is carbon dioxide, has a definite structure. For example, one can imagine the following ecological groupings:

(1) organisms decomposing readily available organic substances ("zymogenic").
(2) Hydrolytics, affecting the resistant biopolymers.
(3) Microflora of dispersion, utilizing low concentrations of hydrolytic products or incomplete oxidation.
(4) Microflora utilizing little-accessible products of microorganisms' metabolism, for example, humates ("autochthonous microflora").
(5) Bacteriolytic microflora, destroying the atrophied cells of micro-organisms.

There are also analogous groupings amoung mycelial organisms (Mirchink and Babyeva, 1981), although studies into their role in the biosphere are inadequate.

The interaction of microorganisms within the framework of a supraspecies system determines its capacity to undergo such reactions that cannot be affected by the system's individual components. The most pronounced example of such interaction is syntrophism (Zavarzin, Bonch-Osmolovskaya, 1981). Syntrophism is most important in cases when the metabolic product utilized by the second component of the syntrophic pair inhibits the metabolism of the producer. Of course, each component of a syntrophic system may be cultivated separately in an artificial medium, but such a pure culture in a sense turns out to be an artifact, though it may be essential for an understanding of the physiology of the organisms. Syntrophic interactions impose an essential limitation on the interpretation of pure culture results, since an organism's metabolism may substantially differ in a syntrophic associa-tion. For the purposes of global microbiology a syntrophic association is considered as a single entity.

The conclusions drawn from the existence of stable microbial systems prove to be of fundamental importance for microbiology. As a matter of fact, the subject of the investigation undergoes change — instead of a single organism there appears a certain organized multiplicity of them. Added to that at the cellular level there is regulation at the intercellular and interspecies levels which has its own means of realization: firstly, the product-substrate interaction of microorganisms; secondly, a concen-trated mechanism for regulating the growth kinetics of microorganisms;

and thirdly, a specific action mechanism by substances of an antibiotic nature and hormone-like activators and growth inhibitors.

In all probability the elective culture remains the most effective method of investigating the organization of microbial systems, whereas an optimal medium is essential for studying the pure culture. A step-by-step break down of the microbial system according to the stages of substrate conversion allows us to determine the basic route in the transport chain. However, there always remains an element of uncertainty, which makes one regard the supraspecies system as something of a single entity.

AN INTEGRAL EVALUATION OF THE ACTIVITY OF MICROBIAL SYSTEMS

Since a microbial system functions as an integrated whole a generalized evaluation of its activity is essential. Such an evaluation cannot be drawn with sufficient completeness and validity from the definition of activity of even the leading components of the system, since such activity is influenced by the interaction within the system's framework. The majority of problems resolved by terrestrial sciences do not require a detailed picture of the microbial system's internal organization. On the other hand, this organization is of special interest to general microbiology. What is important for the majority of interdisciplinary problems of immediate interest is a summary evaluation of the activity of a microbial system and a forecast of its behaviour as an integrated entity. Such an asssessment proves decisive for microbiology too, because it alone makes it possible to establish whether the analysis of the system's organization was correctly conducted.

There is little wonder that at present we see a broad stream of investigations in which the activity of microflora is evaluated by agrochemical, hydrochemical, geophysical and other integral methods. It should be recognized that, unburdened by tradition and disciplinary frameworks, the geochemists' attitude to the need for taking into account the activity of the microflora proved simpler owing to the influence of V. I. Vernadsky, who had fully realized the problem as far back as 1926. At the same time microbiologists who turned to integral evaluation quickly lost faith because of the sceptical attitude of their colleagues in the profession. The unjustified aim of microbiologists to

keep within the established frameworks of their discipline is bound to result in the near future in increasing backwardness.

The traditional approach of ecological microbiology is to compare populations of microorganisms with the observed scope of the process. A vast amount of materials from soil and marine microbiology has been accumulated along these lines. These data should be used to the maximum within the limits allowing for correct interpretation, since they have practically covered all the earth, and the accumulation of data on a new methodological foundation will require years and years of effort and tremendous material expenditure.

Taking account of the specific physiological groups is fraught with a fundamental limitation of the deductive approach, since the elective environment is present. At the same time the modern analytical basis of contemporary field microbiology makes it possible to go over to natural concentrations of substrate. One can easily imagine the artifact created by gram concentrations of substrate wherever microgram concentrations exist in nature. Therefore the afore-mentioned approaches should be regarded as indicators.

The next step is to establish the intensity of the process, for which the isotope method has become the most widespread. The intensity of the process is determined by the activity of the microflora and presents the greatest interest to the microbiologist. Small cycles as a rule emerge in a microbial system, for example, from nitrate reduction and the nitrification of the second phase. The rule of cycle closure operates even in a relatively small microbial system. As a result the total effect turns out to be several times less than the one that stems from the determination of activity.

The total effect is due to the imbalance of small cycles and the withdrawal of substances from the cycle. Moreover, mineral components and humus are subjected to burial and create a basis for successions. For example, the creation of conditions by microorganisms for the formation of clay minerals results in the formation of gas- and water-confining beds and conserves the system. Here the total effect closely coincides with the geological effect which is determined by the concrete geological situation, diagenesis and the formation of minerals. Though this process is outside the framework of the microbial system, it closely influences its fate. At the same time it provides actual grounds for tracing the fate of the microbial system in the past.

The total effect of the microbial system's actions may be determined

with various degrees of generalization and this in equal measure depends both on the methodological approach, and on the methodological task. For example, direct microscopical methods are limited to fractions of the square millimetre, and very sharp differences or much statistical work are needed in order to obtain a statistically significant conclusion. A similar situation prevails, for example, when using gas chromatography to investigate soil air. The one-millilitre sample describes only a small area of a rather heterogeneous system, precluding a statistically significant assessment of the total activity over areas of any size.

On the other hand, rather generalized data on variations of biogenic gases throughout the entire section of the atmosphere are available. For example, the content of methane over the Northern Hemisphere increases in Spring when bodies of water open up, and this could serve to assess the activity of microorganisms over the entire period from the beginning of the freeze-up.

Nearly all the available methods are associated with the extraction of a sample or upsetting the condition of the study object in some other way. Meanwhile for the purpose of global microbiology the object should be preserved intact and information must be obtained from it step by step, as in a lysimeter. As is known, expeditions are conducted at times favouring fieldwork, whereas reliable conclusions require information round the year. It can be obtained either by organizing regular monitoring of the geographic network, or devising some remote method of evaluating microfloral activity.

In all probability, a fundamental solution of the problem could be the elaboration of a method for determining the flow of biogenic gases over considerable areas, i.e., a modification of aerospace methods, most likely based on spectroscopy. There is no question of such an approach influencing the study object and it allows for repeated observations. Determining the flow of biogenic gases from the earth's surface is a fundamental interdisciplinary task. Determining the flow fo nitrous oxide may provide information about the state of the nitrogen cycle, determining the flow of methane about the activity of anaerobic microflora which, as was demonstrated above, is of fundamental importance for the biosphere. Lastly, the flow of oxygen and carbon dioxide in day-time and night-time may indicate the activity of biological processes. These data are important themselves also for forecasting the state of the atmosphere and consequences of cardinal significance for

mankind. It is superfluous to speak of the difficulty of tackling the task, but in the face of global biospheric crises the urgency of posing the task in unquestionable.

In a local area variant the gas exchange of the soil becomes the most important method for assessing total microflora activity. It allows one to fill the gap between the study of microbial populations or communities under controlled conditions of laboratory culture and global recalculations. It is necessary to get a picture of the microbial activity within a single landscape, for which appropriate methods must be elaborated.

An examination of the problems of global microbiology clearly shows the sharp demarcation line between the study of individual cases and a general approach. In the majority of instances individual objects are important only so far as they provide an understanding of the general law-bound regularities. The significance of studying objects for global microbiology is directly dependent on their typicality. The wish to study individual, unique objects-rarities is characteristic of man, but it may create a distorted picture of the processes as a whole, since it is not usually accompanied by a quantitative assessment of importance.

What can global microbiology yield? What can man do, having sorted out the gigantic natural mechanisms? Could we change their course? It is clearly best to foresee the future during the first stage, in order to avoid those unfavourable consequences that *can* be avoided.

REFERENCES

Bach W. *et al.* (1980). The Carbon Dioxide Problem. An Interdisciplinary Survey. *Experientia*, **36**, No. 767, p. 890.
Budyko M. I. (1980). *Climate, Its Past and Future.* Gosmeteoizdat Publishers, Leningrad, (in Russian).
Chan Y, H., Olson J. S. (1960). 'Limits on the Organic Storage of Carbon From Burning Fossil Fuels.' *J. Environment. Manag.* **11**, No. 147, p. 163.
Hahn J., Junge C. (1977). 'Atmospheric Nitrous Oxide: A Critical Review.' *Ztschr. Naturforsch* **32a**, No. 190, p. 214.
Kondratyeva E. N., Gogotov I. N. (1980). *Molecular Hydrogen in the Metabolism of Microorganisms.* Nauka Publishers, Moscow (in Russian).
Konings W. H., Veldkamp H. V. (1980). 'Phenotypic Responses to Environmental Change.' In *Contemporary Microbial Ecology.* London, p. 161.
Mirchink T. G., Babyeva I. P. (1981) 'Mycelium Forming Fungi and Yeasts in Natural Ecosystems.' In *Zhurn. Obshch. Biol.*, **42**, No. 3, p. 390 (in Russian).
Pechürkin N. S. (1978). *Population Microbiology.* Nauka Publishers, Novosibirsk (in Russian).

Rosswall, T. (1978). 'Impact of Massive Microbe Mediated Transformations on the Global Environment Microbial Activity Affecting the Thickness of the Ozone Layer and the CO_2 Concentration in the Atomsphere.' *Rept. XII Int. Cong. Microbiol.* Munchen.

SCOPE—13 (1980). The Global Carbon Cycle (Éds. Bolin B., Degens E. T., Kempe S., Ketner P.) J. Wiley, New York.

Stater J. H., Goodwin D. (1980). 'Microbial Adaptation and Selection. In *Contemporary Microbial Ecology.* Academic Press, London.

Talroze V. L., Larin I. K., Poroikova A. I. (1979). 'Chemical Reactions of Microbiogenic Gases in Terrestrial Atomsphere.' In *The Role of Microorganisms in the Cycle of Gases in Nature,* Nauka Publishers Moscow, p. 35 (in Russian).

Zavarzin G. A. (1979). 'Microorganisms and the Composition of the Atmosphere.' In: *The Role of Microorganisms in the Cycle of Gas in Nature.* Nauka Publishers, Moscow (in Russian).

Zavarzin G. A., Bonch-Osmolovskaya E. A. (1981). 'Synthrophic Interactions in Communities of Microorganisms.' In *Izv. AN SSSR Biological Series,* No. 2, p. 165 (in Russian).

Institute of Microbiology,
U.S.S.R. Academy of Sciences,
Puschino,
U.S.S.R.

C.-G. Hedén

The Scale-Factor Paradox in Biotechnology

INTRODUCTION

As we all know, economic growth had dominated all planning processes in the industrialized parts of the world for so long that the recent slow-down came as a very painful experience indeed.

Noting that "Big Government", "Big Labour" and "Big Business" were part of the problem (militarization, inflation, unemployment, environmental and social stresses, etc.), many people expressed doubts about the ability of such organizations to provide the solution. This, they claimed, could instead be found under the banner of Schumacher's motto: "Small is beautiful" ([1]).

The consequent polarization between big and small is unfortunate, because it obscures a fact that ought to be obvious, namely that *both* approaches are needed, if we want to avoid traps like protectionism and increased bureaucracy.

Sure enough, such moves may offer short-term solutions, as long as we believe in the merits of abstractions like "national sovereignty", GNP as a measure of development and science and technology as secondary factors in the economy. However, if we accept the "one world" concept, and the fact that products of the mind cannot be contained within national borderlines as easily as products of industry, we are inevitably led to take a long-range strategic look at the impact of "high-technology" on the global matrix of resource endowments. This immediately brings such resources as our renewable energy base and the manpower distribution into focus.

WHAT IS HIGH-TECHNOLOGY?

Modern biotechnology is often referred to as "high-tech", which makes us associate it with *capital-intensive* systems like broad-band communication systems, automated factories and artificial organs. This makes us

151

forget that its fundamental characteristic is *knowledge intensity*, and that the economics of the innovations concerned must be expressed, either as the cost per bit of information and unit of product handled, or as the value of the gains made: in time, in energy or in life-quality.

Viewed in this way, high-tech is normally not only *cheap* but as a rule, it is actually also *simple to use* — if measured in relation to performance. This is true, not for what I like to call the the King of high-tech: Microelectronics, but also for its Queen: Biotechnology.

The low price and energy efficiency of transistor radios is impressive, and one certainly doesn't need to know anything about solid-state physics to use them. Similarly the efficiency of a microbial starter culture is impressive, and one doesn't need to know any molecular genetics to use it.

THE SCALE FACTOR IN HIGH TECHNOLOGY

When discussing the scale-factor in relation to high-tech I think that it is essential to make a distinction between the economy of scale in *primary production* and in *secondary manufacturing*. The production of electronic components (microprocessors, memory devices, etc.) and of biocatalysts (enzymes and cells) requires large and advanced facilities.

However, the devices and the processes of which such items are part, may well have their major impact in small and rather unsophisticated environments.

Nobody disputes that it is economically sound to make "chips", as well as baker's yeast, in big and largely automated factories, but as surely as a transistor radio can generate knowledge in a primitive village, a package of lyophilized yeast cells can to-day produce bread or beer on a small scale in a very modest environment. Tomorrow it will — with the aid of genetic engineering and enzyme technology — give us the basis for a new kind of "cottage industry".

One would assume that the planners of the infrastructures that are needed to serve those two operational levels would have drawn the consequences of such rather self-evident observations. However, when I compare the large national programmes for microelectronics and genetic engineering, which now emerge in many countries, on the one hand, with the efforts that will support their implementation, on the other (software development, education in applied science etc.), I get a feeling that it is now the highly visible, centralized, initiatives that have

the greatest political "sex appeal". The same is also true ([2]) for the very expensive applications of biotechnology in the pharmaceutical industry (Table I).

In fact, decision-makers often seem so fascinated by the market estimates for hormones, interferons, and various other drugs (Table II), that they tend to forget the long-range, global significance of genetic engineering and biotechnology.

In my opinion, this significance resides in the potential of those two fields for a synergistic interaction with two other areas, namely with

TABLE I
Some current efforts in the pharmaceutical industry

	Million US dollars
Overall spending on "New Biotechnology" (Merck)	338
Alpha interferon plant (Searle)	106
Development of beta-interferon (CETUS/Shell Oil)	40
Plant for beta-interferon (CETUS/Shell Oil)	10
Clinical trials with gamma-interferon (Biogene)	50
Clinical trials with human insulin (Lilly)	12—20
Insulin plant (Lilly)	60

TABLE II
Some market estimates for pharmaceutical applications of genetic engineering

		US dollars	
Total world pharmaceutical market (1981)		76.3	Bill.
World market for pharmaceuticals by "New Biotechnology" in the year 2000		5—10	Bill.
Anti-arthritic peptide		500	Mill.
Insulin		400	Mill.
Monoclonal diagnostic kits	1982	20	Mill.
	1990	300/500	Mill.
Monoclonal therapeutics	1982	0	Mill.
	1990	500/1000	Mill.
Human vaccine industry	1980	95	Mill.
	1985	275	Mill.
Veterinary vaccine industry 1981 (Growth 20—25% per annum)		1	Bill.
Interferons	1981	50	Mill.
	if highly effective perhaps	8.25	Bill.

microelectronics and communication technology. The reason for this is simply that the latter two areas also support decentralization. And this, in my view, is probably the ultimate solution that will hopefully help us to eliminate structural unemployment and offset some of the social costs (diseases, shantytowns and crime) as well as much of the energy expenditures (transportation, centralized services) that are associated with urbanization.

A quick look (Table III) at the demographic data from large parts of the world indicates that urbanization is in fact now getting out of hand ([3]). However, there is little doubt in my mind that ecological imperatives and the drive towards self-reliance at all levels (individual, community, nation) will move us in the direction of decentralization.

TABLE III

Estimates and rough projections of populations of selected urban agglomerations in developing countries (millions of persons)

	1960	1970	1975	2000
Calcutta	5.5	6.9	8.1	19.7
Mexico City	4.9	8.6	10.9	31.6
Greater Bombay	4.1	5.8	7.1	19.1
Greater Cairo	3.7	5.7	6.9	16.4
Jakarta	2.7	4.3	5.6	16.9
Seoul	2.4	5.4	7.3	18.7
Delhi	2.3	3.5	4.5	13.2
Manila	2.2	3.5	4.4	12.7
Teheran	1.9	3.4	4.4	13.8
Karachi	1.8	3.3	4.5	15.9
Bogota	1.7	2.6	3.4	9.5
Lagos	0.8	1.4	2.1	9.4

I regard this as an optimistic outlook and hope for a rapid growth of this decentralization trend, in spite of a natural opposition from the guardians of outdated technologies, as well as from the advocates for centralism and hierarchical decision-making.

THE POST-INDUSTRIAL SOCIETY

When we consider the post-industrial society as a hybrid between a "Biosociety" and an "Information Society", it is easy to visualize the

post-industrial world as a network of service-oriented settlements. Such settlements would, of course, be supplemented with manufacturing enclaves for the various *mature* products (such as transport and communication equipment) that will be needed to satisfy basic human needs. We can also assume that they would use systems integration and energy-cascading to ensure an optimal utilization of non-renewable resources. However, such manufacturing enclaves would be served by data-links and geared to robots for laser cutting and other advanced operations, that are suitable for computer-aided design and manufacturing (CAD/CAM). Consequently, they would certainly not require much manpower. The inventions required for an implementation of this scenario already exist, but the social innovations that would transform its inevitable consequence — *structural unemployment* — from being a problem to becoming an opportunity, I think appear much too slowly. What is for instance the content of the biology teaching that we use to prepare our children for entering a biosociety with lots and lots of spare time? Do we really use biology as a treasure chest for stimulating young, inquisitive and creative minds?

Perhaps this is an area where the adaptations will be easiest for the countries that entered the industrialization phase late, and now pass through it so quickly that the traditional value patterns have not had time to erode. Is it not, for instance, reasonable to assume that a genetics-mediated transition into a "Biosociety" will be easier for those countries in Asia, that have a long tradition in food fermentation and biorecycling, than it will be for the countries that once built their industries on the basis of cheap energy and imported raw materials? And is it not probable that it will be easier to enter the "Information Society" directly, with a broad-band communication system, than if thousands of tonnes of copper wire and mountains of mechanical relays must first be written off?

In fact, the United Nations Industrial Development Organization (UNIDO) has often stressed that there are certain "high technologies" which developing countries can now ill afford to neglect. Interestingly enough, those technologies may also be precisely those which the rich world should make every effort to transfer to poor countries, if the outstanding debts of the latter are ever to be serviced. I definitely share the view that such technologies merit the name "Technologies for Humanity" ([4]). They are much too important to become instruments for international competition rather than for cooperation.

It is true that the mid-sixties, which have been called "the era of technological transplantation" ([5]), taught us that Northern Technology does not go very far towards solving such problems as under- and unemployment, low productivity, urbanization and a negative balance of payment.

However, this can hardly be used as an argument against *all* Northern Technologies. After all, effective associated programmes, aimed at inducing innovation acceptance, are fairly recent. Also, we have to admit that too much of the official development assistance was earlier focused on fairly small segments of population.

Hopefully, we thought, the benefits to rich landowners and manufacturers would "trickle down" to the poor. By now we have learnt that this process is much too slow, and that we should look towards areas like genetic engineering and biotechnology, for the very reason that they can have a highly decentralized impact. And this is an impact which has a potential for causing catalytic effects throughout the whole global economy — a prerequisite for the major assistance efforts which I think must now be considered.

BIOTECHNOLOGY TRANSFER

The need for building an infrastructure for applied microbiology and biotechnology in developing countries was noted by applied microbiologists long before the energy crisis, and it has actually been a central theme for the UNEP/UNESCO/ICRO Microbiology Panel for the past two decades.

In recent years the World Academy of Art and Science (WAAS) has also concerned itself with the problem of technology transfer in this field. It is, for instance, recently considered initiating an exploratory study on the possibility of establishing a self-supporting "Biological Resource Development Corporation", based on two feasibility studies concerned with defining the design criteria for mobile pilot plants built for pretreatment and for fermentation of lignocellulose-derived substrates.

Actually, the Microbiology Panel as early as 1965 saw the need for training and research centres in developing countries, and it considered many approaches, including a proposal from my own laboratory ([6]). This involved the establishment of fact-finding teams of bioengineers having access to mobile and highly versatile field laboratories. Those

"Biological Resource Development Teams" (BIOREDs) would not only evaluate the local nutritional situation and the existing microbiological problems and processes. They were also supposed to carry out experiments on fermentative upgrading and to have the capacity to produce various inoculants and starters on a fairly large scale (Fig. 1). The relation between the mobile field station and the base laboratories was discussed in detail, but in spite of strong support from many other international organizations, the time was not ripe twenty years ago to raise the funds needed to support mobile fermentation facilities.

However, a worldwide network of Microbiological Resource Centres (MIRCENs) was established, and this now stands ready, when flexible small-scale facilities eventually seem ready to materialize, in order to meet expected increases in the world's protein demand.

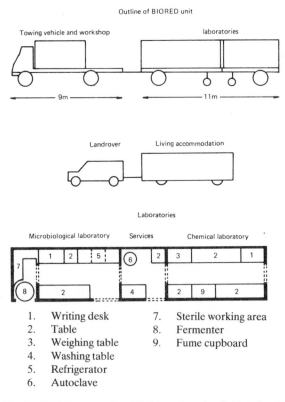

1. Writing desk
2. Table
3. Weighing table
4. Washing table
5. Refrigerator
6. Autoclave
7. Sterile working area
8. Fermenter
9. Fume cupboard

Fig. 1. 1965 Concept of mobile bioengineering field station (6).

SCP AND THE SCALE-FACTOR PARADOX

Some years ago Een ([7]) pointed out that the food industry was likely to pass through a scale-factor inflexion point. The fermentation industry is, of course, exposed to the same forces (Table IV), and it is interesting to note that some of the corporations that have gained considerable experience in exploiting the advantages of large-scale protein production (SCP) are now directing attention to the opportunities offered by small-scale operation (Table V).

Yeast production and brewing are classical examples of the economy of scale in fermentation, and large-scale production of SCP (10 000 – 100 000 tonnes/annum) from n-paraffins and methanol is also economi-

TABLE IV

Factors contributing to the contraction of the food industry
unit size

Trend towards increased national and regional self-reliance
Reduced vulnerability both of society and of company
Diminishing world trade
Rising transport costs
Recycling of waste to agriculture
Legislation regarding environmental protection
Increased use of decentralized energy sources
Increased demands for job satisfaction
Cheaper and more reliable automation
Cheaper and better telecommunications
Cheaper and better information storage and processing

TABLE V

Advantages in various types of SCP production

Large scale	Small scale
Low labour costs	Employment generation
Efficient energy recovery	Modest energy needs
Effective control of process safety and product quality	Process flexibility
Substrates available on attractive commercial terms	Utilization of locally produced raw materials.
Simplified storage and marketing	Suitable for immediate consumption and compatible with local needs

cally feasible under certain circumstances (cf. the British Petroleum and the Imperial Chemical Industries processes).

However, as I indicated, small-scale operations are now starting to attract attention from big business. The German firms Hoechst and UHDE ([8]) have for instance recently developed a mobile and quite robust paddle-wheel fermenter for making protein-rich fodder in the 100−1000 tonnes/annum range (Fig. 2). The processes developed are intended for molasses, sugar juices, fruit syrups, coconut milk, whey and starch, and they can be run wherever a basic infrastructure exists (water, energy, personnel).

The costs are cut in various ways, for instance by avoiding an expensive drying process. This can be achieved if an adequate number

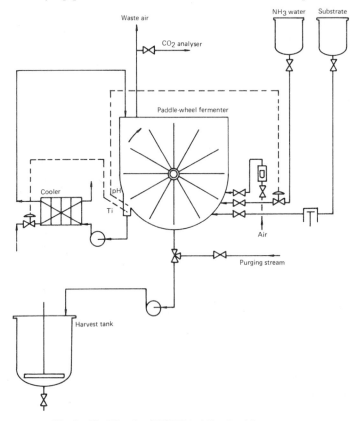

Fig. 2. The Hoechst/UHDE paddle-wheel fermenter.

of cattle, pigs or poultry is available to be fed with a fresh, semisolid preparation.

The pretreatment, of course, varies with the substrate, but it is based on standard practices adapted to the capacity of the fermenter (40 cubic metres filled with about 12 cubic metres of culture) which operates in a "feed-batch" mode. Major savings are also achieved by choosing fermentation processes that are sufficiently resistant to obviate the need for sterile operation, and by aerating with a fan rather than with a costly compressor. This is made possible by the slowly rotating, large-scale agitator which also gives good stirring without much shear and foam formation.

Tate and Lyle in the U. K. ([9]) noted early on the need for upgrading agricultural waste with simple equipment and have successfully operated a tower reactor in Belize. This was made from fibreglass-reinforced polypropylene, and it could be operated non-aseptically with moulds that tolerated a high temperature and a low pH. Such a unit, aimed at a capacity of 500—1500 tonnes of SCP per annum, not only demonstrated the crucial importance of village-level, on-site training. It also later provided some of the data which Lewis in the U.K. could use in a revealing energy analysis ([10]) that was performed to evaluate the impact of applied microbiology on developing countries ([11]). This study, which is summarized in Table VI, goes a long way to explain why *both* large-scale and small-scale approaches can be defended. And this, of course, goes very well with the prediction that the two approaches will supplement each other.

In such a scenario very large central units would produce the genetically modified, often mixed, starter cultures that would then be used in a large number of small, non-aseptic fermenters, located next to the source of the feedstock.

SCALE-FACTOR ASPECTS ON R&D IN BIOTECHNOLOGY

When looking at the dollar figures in Table I, which illustrated the current efforts of the pharmaceutical industry in the fields of "new biotechnology", one might get a feeling that this area is just for "the Big Boys". This is certainly true for the major human drugs, but even here there are some speciality fields, like peptide hormones, where the development cost might not go beyond 5—10 million US dollars ([12]).

However, even such figures, which reflect very high costs for patent-

TABLE VI
Input/output estimates for SCP production

A. Physical input

(mJ/kg SCP)

Substrate

Operational	Methane	Methanol	Paraffin	Molasses	Solid agric. waste	Agricult. effluent
Substrate	102.75	84.40	45.27	16.23	17.44	0
Chemicals	13.07	13.07	13.84	0.38	5.47	5.47
Water	0.03	0.02	0.13	0.54	0.08	0.08
Electricity	14.00	18.62	31.50	14.00	21.00	11.20
Fuel oil	—	—	25.15	6.48	—	—
Capital						
Stainless steel	0.26	0.26	0.33	0.20	—	—
Structural steel	0.30	0.30	0.35	0.20	0	0
Cement	0.08	0.08	0.10	0.07	0	0
Polypropylene	—	—	—	—	0.07	0.07
total	130	117	117	48	44	17

B. Yields and labour requirements

	Methane	Methanol	Paraffin	Molasses	Solid agric. waste	Agricult. effluent
Output tonnes/year	50000	50000	100000	50000	500	1500
Area used (ha)						
production site	12	12	20	8	0.1	0.1
cultivation	—	—	—	18000	200	720
Yield (kg/ha/year)	4167 $\times 10^3$	4167 $\times 10^3$	5000 $\times 10^3$	2800	2500	2100
Energy subsidy (GJ/ha/year)	541710	487539	585000	134	110	36
Manpower need	200	200	375	200	10	40
Tonnes protein/man/year	200	200	170	160	30	20

ing, testing and scale-up, tend to make many people believe that this type of high tech is *always* also high-cost. For one thing this makes one forget the cost-advantages of shared facilities (see below), but more importantly, the high costs of human drug development might scare away some scientists in developing countries from the veterinary (vaccines and hormones), agricultural (biofuels and bio-insecticides).

and environmental applications (biorecycling and detoxification). Those are fields where rapid progress may be attained even without the use of the much publicized DNA-hybrid methods. But not even such methods − that are now routine in laboratories all over the world − are particularly costly. Actually, even the most rapidly growing segment of "the new biotechnology", the monoclonal antibody field, can yield new diagnostic kits or therapeutic approaches (drug-targeting and imaging) in facilities, the costs of which would certainly not impress a modern physicist (Table VII).

TABLE VII
Cost estimate for a monoclonal antibody laboratory

Item	Cost (US dollars)
CO_2 incubator	3000− 5000
Tissue culture sterile hood	3000− 8000
Inverted microscope	1000− 8000
Standard microscope	1000− 3000
Water bath	300
Centrifuge	2000− 8000
Autoclave	1500−15000
pH-meter	400− 1000
Balance	1200
Liquid nitrogen storage containers	3000− 6000
Water distillation system	6000− 8000
Mouse cages	3000
Purchase of 50 mice for breeding stock	200
Maintain 150 mice for 12 months at $ 0.035 per day and mouse	1890
Tissue culture flasks and plates	4000
Tissue culture medium, buffers, antibiotics	2000
Foetal calf serum	2000
Plates for assay of antibiotics	2000
Additional glassware (beakers, bottles, pipettes)	1000
Total estimated cost	41000−75000

This represents initial costs: labour costs and the laboratory facility itself are not included. Estimated costs for optional equipment are:

Fluorescent microscope	$ 1500
Scintillation counter	$ 15000− 25000
Flow cytometer	$ 125000−250000
Total estimated cost with added equipment	$ 180000−353000

Source: US Natl. Res. Council *op. cit* p. 204.

THE SCOPE FOR INTERNATIONAL COOPERATION

As mentioned earlier there exists a number of Microbiological Resource Centres. However, none of them has all the resources needed to realize the full impact of the many recent breakthroughs in fermentation, enzyme technology and genetic engineering on the special problems of developing countries.

From this point of view the decision of 5 April 1984 by thirteen governments to establish an International Centre for Genetic Engineering and Biotechnology in Trieste/New Delhi is not only a historic event, but also an acid test for the major industrialized countries of the world. They can of course choose to ignore it, and continue using science and technology as instruments for foreign policy, or they can use it as a powerful switchboard for free communication between the world's applied microbiologists, microbial geneticists and bioengineers.

Their choice will demonstrate if decision-makers fully realize the international implications of the Biosociety-concept. Either they settle for full-scale information exchange and research cooperation on bio-resource utilization for the five Fs: food, fodder, fuel, fertilizer and fibre; or they will find themselves unprepared for the social consequences when they open that Pandora's box which is part of the technology of the 1990s: macromolecular engineering. This area is the mix of bioinformatics, quantum biochemistry and computer graphics which makes it possible to design macromolecules (vaccines as well as toxins) on a video-screen before a laboratory for developing a production method is chosen. This is likely to be a laboratory for bioengineering, but I have to admit that it might very well also be a laboratory for chemical synthesis.

I make this my concluding remark in order to stress that we must watch out for the dangers of a disciplinary focus, whenever we try to look into the future.

If anything, the developments in biotechnology over the last decade demonstrate that this type of high technology grows best at the interfaces between our traditional academic fields. This indicates one of the big problems that science policy faces today.

REFERENCES

[1]   Schumacher, E. F. (1973). *"Small Is Beautiful"*. Harper and Row, London.
[2]   'Biotechnology and the Developing Countries: Applications for the Pharmaceu-

tical Industry and Agriculture.' UNIDO/IS. 452. 12. March 1984. Prepared by the UNIDO Secretariat.

[3]   'The Global 2000 Report to the President.' A report by the Council on Environmental Quality and the US Dept. of State. Study Director G. O. Barney. Vol. 1. US Govt. Printing Office 1981. O−344−113 QL 3.

[4]   Preparatory documents for High-level symposium planned in preparation of UNIDO IV (Vienna, May 24−26, 1984). Vienna Institute for Development. A−1010 Vienna, Austria.

[5]   Wiener, A. (1984). 'The Linkage Between Technological and Social Innovation and the Promotion of Social Innovation Through International Prizes.' In *Social Innovations for Development.* Eds. C.-G. Hedén and A. King. Pergamon Press, New York.

[6]   'BIORED-Biological Resource Development Teams'. Design and feasibility studies prepared by C.-G. Hedén, R. Brookes and B. Sikyta for the UNESCO/ICRO Microbiology Panel, the Swedish IBP-committee, the Advisory Council of IAMS and the Roy. Swed. Acad. Eng. Sciences. MIRCEN/Stockholm. Dec. 15th 1965.

[7]   Een, G. (1979), 'Food Industry in the Year 2000.' Paper at Workshop: Size and Productive Efficiency − the Wider Implications. IIASA. A. 2361. Laxenburg. Vienna. 25−29, 1979.

[8]   Uhde and Hoechst. (1984). 'Mini-Process for the Fermentation of Organic Raw Materials.' Undated pamphlet (Uhde GmbH, Degging-strasse 10−12. Postfach 262.4600. Dortmund 1. FRG).

[9]   Imrie, F. K. E. and Righelato, R. C. (1976). In *Food from Waste.* Eds. G. G. Birch, K. J. Parker and J. T. Worgan. Applied Science Publishers, London. p. 79.

[10] Lewis, C. W. (1976). *J. Appl. Chem. Biotech.* **26**, 568

[11] Slesser, M., Lewis, C., and I. Hounam (1981). 'The Potential for Application of Microbiology Within Third World Rural Communities.' Undated booklet compiled with financial assistance of UNESCO (100 Montrose St. Glasgow, G 4 OLZ).

[12] Editorial. 'The new peptide drugs', *Chemical Week.* **133, 28** Sept. 1983. p. 40.

UNEP/UNESCO/ICRO Microbiological Resources Centre (MIRCEN),
Karolinska Institute,
10401 Stockholm,
Sweden.

H. G. Gyllenberg and E. J. DaSilva

Biotechnological Considerations in Research and Development

INTRODUCTION

A recently published report ([4]) lists no less than ten different definitions of biotechnology put forward during the last few years. The authors of the report suggest themselves the following:

[Biotechnology is] the application of scientific and engineering principles to the processing of materials by biological agents to provide goods and services.

In this otherwise clear definition the concept "biological agents" remains undefined. However, a working group of the Finnish Ministry of Education (5) in accepting the definition, explains what biological agents are. The amended text then reads:

Biotechnology is the application of scientific and engineering principles to the processing of materials by biological agents (living cells and enzymes) to provide goods and services.

In the text that follows the concept of biotechnology is used in this sense.

In recent years a vast number of reports on, and recommendations for, national and international policies in the field of biotechnology have been published. This review tries to present the most important points that have been stressed and further discussed. However, the available background material does not represent a complete coverage — neither geographically nor by discipline. Easily accessible information relates to the "high technology" of the industrialized countries and to those applications that are expected to be attractive to these societies.

In spite of this, a discussion related to the developing countries is included. However, since direct contributions to such a discussion from the side of the developing countries themselves are not available, the views put forward may reflect sources such as the UN and the national development agencies or their representatives.

Science and technology policies deal with the identification of potentials and the interpretation of the outputs of this identification process into prospects. In this review an analysis will be made of present potentials in biotechnology. The scope is wide, stretching from the purification of sewage to the production of enzymes for diagnostic use. Therefore, a complete coverage is not aimed at, and a comparative evaluation of different potentials is avoided. This is certainly a weakness, but sound evaluation and comparison cannot be founded on the information presently available.

Good policies lead to prospects realized. The following discussion therefore restricts itself to more or less generally accepted options and objectives; "science fiction" is rejected. Again, although absolute coverage cannot be reached, a broad discussion of existing possibilities has been the goal. When discussing prospects, realized or realizable, impacts on society become important. Therefore, this context reflects socio-economic effects as well as those ethical problems that the breakthrough of biotechnology may give rise to.

Biotechnology is not an end of itself. It serves economical goals and technological markets, and concerns ethical values. In this respect biotechnology, on the one hand supports prevailing thoughts and views, but on the other hand provides approaches and perhaps even solutions to new kinds of problems opening new dimensions of human knowledge.

THE TECHNOLOGICAL AIM AND AN EXAMPLE

Policies, potentials and prospects are relevant only in the dimension of progress. Policies in the field of biotechnology, as well as attempts to identify potentials and to design prospects, all aim at progress.

In the context of biological evolution, progress reflects increase and improvement. With constant advances in science, an increase of knowledge and information become more compatible. The laws of biological regulation are, however, also valid: an exponential increase is always levelled-off, a rapid increase comes always to a standstill.

A recent analysis (6) concluded that technology involves a process and a product (produced by that process), and furthermore proposed that progress is the invention of products that are better (cheaper, more useful, etc.) than a previous product that fitted the same purpose. However, it happens — and that is in fact the rule — that scientific and

technological attempts are limited to improving the process, whereas the search for replacing products is overlooked. The process-oriented approach indicates levelling-off because knowledge and information are not substantially increased.

The technological aim then is new products for known purposes. An analysis of policies, potentials and also prospects in biotechnology is, therefore, most closely related to a search for *new products.*

In the last few years biological nitrogen fixation has become a key issue in biotechnology, as well as in policies, potentials and prospects related to biotechnology. The reason for the "renaissance" of interest in biological nitrogen-fixation is obvious. Chemical nitrogen fixation needed for the production of chemical fertilizers is very energy-consuming and as the price of energy increases, the use of chemical fertilizers becomes continually more expensive. Biological nitrogen-fixation also requires considerable amounts of energy, but the "trick" is that the most efficient biological nitrogen-fixing systems are able to rely on "priceless" solar energy. The main goal of the present research on biological nitrogen-fixation as expressed in the Unesco/UNEP MIRCEN Network programme, and in most national programmes in industrialized countries, is obviously a process improvement programme: more efficient symbiotic strains of *Rhizobia,* more successful inoculation techniques, and better agricultural routines relating both to soil preparation and harvest.

Biotechnology in this field, and research in biotechnology generally are hence, according to the "technological aim", in a levelling-off stage. Very much more cannot be achieved. A real breakthrough can be expected only if and when a completely new product is invented. What could then be a new product of biological nitrogen-fixation? Possibly a nitrogen-fixing plant choroplast, which may not be obtainable in the next five or ten years, nor perhaps, in the next fifty. The science and technology policy problem would then be as to whether the process, although in the stage of levelling-off, or the new product in its very early lag phase, should be given priority in the allocation of research funds.

No consistent deduction can be arrived at. Improvement of the already well-known process can still give convincing outputs, especially in developing countries, where it has not been efficiently applied until recent years. In the industrialized countries such results seem to remain insignificant. Probably an acceptable conclusion would be to improve the presently-known process up to its ultimate limits, keeping in mind

that considerable progress can only be achieved by a new breakthrough based on a new scientific idea: a new product, and a new process to produce it.

THE ELEMENTS OF POLICIES, POTENTIALS AND PROSPECTS

Policies

Recent national reviews and programmes in the field of biotechnology have stressed various issues. Some common elements are, however, obvious, e.g. (i) an analysis of the useful impacts of biotechnology, (ii) governmental steering, and (iii) socio-economic consequences.

The above-mentioned technological aim is valid also as far as biotechnology is concerned. Biotechnology, like other branches of technology, is relevant only in the dimension of progress, with perhaps somewhat different emphasis both in the developing and developed countries. Limits are defined by the small scale of domestic markets, and, on the other hand, by the risks connected with bulk-products on the international markets. The best goal would be to identify marginal products which compete well on marginal, yet large enough markets.

Another problem is "know-how" which can either be bought or generated. Buying know-how implies a delay in getting access to the most advanced knowledge; this delay usually involves years, rather than months or weeks. Years are too much in biotechnology, which is know-how-intensive, both as to soft- and hardware. Buying of know-how consists of similar "immaterial" values, it implies an equal "footing" where both parties can define their conditions for buying and selling.

Biotechnology is made of a large area of activities; individual scientists, enterprises and even countries with defined specializations. Selection of specializations requires a good knowledge of the field of biotechnology. Consequently, broader overall direction of specializations requires an educational input.

The educational system reacts, however, only rather slowly to change in the production system. A concertation of education is difficult to achieve because "biotechnological education" involves a considerable number of different and mutually independent educational curricula. The collection of such curricula in a single university (or "centre of excellence") may be difficult and hardly advisable, because of the inherent fact that biotechnology may need specialists of several different

backgrounds. The solution to the education question in biotechnology would seem to lie in promoting a broadly based training mechanism wherein courses and formal education programmes from different universities and colleges can be combined. Successful students from developing countries should seek their post-graduate education at the advanced biotechnology centres of the more wealthy countries. This requires a network of contacts and interaction, such as has already been established by the Unesco-UNEP MIRCEN network. This network constitutes a new model of useful cooperation between developed and developing countries that secures competent and permanent counter-parts in the industrialized countries to corresponding centres in the developing countries.The MIRCEN model is worth further considera-tion and development.

Science and technology are internationally oriented. National isola-tion is not possible in the present world. As far as biotechnology is con-cerned a special feature is that more of the basic theoretical research is performed through direct funding by commercial enterprises. This has al-ready given rise to a situation where even very general knowledge may be classified as secret (pp. 165—166). Such a practice retards dispersion of new knowledge, and increases the need for overlapping basic research: even very basic facts have to be discovered independently in different laboratories. A key task in biotechnology is to overcome this obstacle.

Patents and patent documents represent, by tradition, a useful and efficient system for the transfer of information and knowledge. How-ever, it would be rather strange to have to use this system for the transfer of basic knowledge which, again by tradition, should be freely accessible to every scientist. The problem of knowledge as "material property" is discussed in somewhat more detail later.

It seems that existing patent law, which is being improved on a common uniform basis in the industrialized countries, is not easily applied to biotechnology. The difficulty has been further accentuated by the rapid development of genetic engineering. There is an ever-increasing number of applications for the patenting of DNA-sequences, plasmids, vectors, recombinant DNA, etc., as well as methods to produce such entitites, but there does not seem to exist an international understanding of the underlying conditions and principles. A common international agreement that draws upon the current practices and patented technologies is urgently needed.

Also, current patent procedures are quite slow, and when a patent is

finally obtained the time of protection remains short. This implies that the advantage gained from a patent is not necessarily in relation to the expenditure and other drawbacks associated with a patent application. On the other hand, abstention from applying for patents intensifies the need for secrecy, and imposes further restrictions on the dispersion of information.

A commonly heard claim is that we are on the threshold of an information society. Rightly or wrongly, the new approach to biotechnology, particularly stressed by the breakthroughs in genetic engineering, emphasizes the crucial nature of access to biological information. Recognition of the "race for biological information" has created "*biological informatics*" which comprise those databases and registers that contain biological information in various forms. Development of such databases and registers and the promotion of easy and practical access to them is an area for international co-operation. Although this will require "high technology" applications it is to the advantage of developing countries to participate in networks that guarantee them access to the newest and relevant information.

Biotechnology is a new challenge to the field of science and technology policy. The responsibility is shared by both the public and the private sector, i.e. R & D is performed in universities and governmental research institutes but also by relevant industrial enterprises. Governmental impact is primarily shown by resource allocation. Since biotechnology is "know-how-intensive", success and progress are decisively dependent on the quantity and quality of the allocations to biotechnology R & D.

Allocating "respectable" resources is difficult, due to the fact that distinct priority areas are difficult to set in biotechnology. This may sound contradictory to what was said earlier, but in essence no contradiction exists. Specializations are the result of the prevailing conditions in the various countries, but within the "specializations" it is, as the cited example of biological nitrogen-fixation above may have already shown, almost impossible to define where the allocations should be concentrated. The dilemma of giving priority to process improvement or to new replacement products will not be easily resolved.

Governmental resource allocation cannot, however, be left only to intuition. Possibly governmental allocations should promote risk-taking and avoid dogmatism. Governmental allocations should also aim at an increased scope and perspective and at guaranteeing necessary con-

tinuity. Another governmental approach could be the improvement of the management of large projects, since, in the field of biotechnology, research projects also tend to grow larger. Scientific managers of such projects still have only "trial and error" training in management. This implies suboptimal efficiency and hence lost resources. Of course, science is not business, and cost-benefit criteria can hardly be applied to scientific work in the same way as in industry; nevertheless, it is a fact that some research groups are more efficient than others (7). The problem also in biotechnology is to increase scientific efficiency.

Government involvement is needed also to deal with international matters. The problems related to the patents are one case. Another concerns the uniformity of testing procedures needed to get new drugs and related materials accepted in various countries. At present, acceptance in one country is not normally valid in another, and this may result in further expenditure. Uniform legislation and an international understanding to obtain one accepted procedure facilitates acquisition of general approval and indicates considerable progress. Such an understanding requires, however, governmental involvement in the countries concerned.

Consistent breakthroughs in biotechnology require acceptance from the public sector. Where public trust is lacking, suspicions concerning failing security and exaggerated latent health risks abound. Governments have to either show that existing regulations provide the necessary basis for avoiding health risks or to implement new regulations. It is obvious that all regulations should concern not only laboratory conditions, but also the conditions pertaining to large-scale industrial production. Biotechnology express itself in society through its products, its values, as well as various activities ([8]). These expressions are probably worldwide. Phenomena first encountered in big countries are rapidly transferred to smaller communities. For example, concerns and doubts relating to the potential health hazards of genetic research were first voiced in the technically-advanced societies in the early 1970s. By the end of that decade, similar expressions were encountered in the developing countries as expertise in such research developed. Indeed, future development may not differ considerably from the experiences accumulated from the breakthrough interested in the field of electronics.

In the framework of policy issues it has to be realized that the "reflections" of biotechnology will be both material and immaterial.

Immaterial is the philosophical question of how to define living and dead material. A DNA-sequence is dead when isolated outside the living cell or when artificially synthesized, but it is living when it is reproduced within a living system. Whether the DNA-sequence is living or dead, is by no means a rhethorical question. It is quite valid because the answer may influence coming generations in their understanding of nature and of the world itself.

Furthermore, ethical "barriers" are soon to be faced. How far can man manipulate other living organisms? This question does not concern security of risks. It is a purely ethical question related to the problem of euthanasis. Again, the answer to this question may probably influence the thinking of future generations.

The problem of "who owns knowledge" has been referred to earlier. This is a relevant question in a situation where fundamental findings are also patented in the field of biotechnology. The problem has been thoroughly dealt with elsewhere ([9], [10]) and needs no further elaboration here. However, touching upon this matter in the context of governmental policies in relation to biotechnology, it is worthwhile to note that secrecy always has its limits: it tends to protect existing knowledge which comes from the free exchange and free testing of new ideas, and is the most significant incentive for social and economic development.

Turning from immaterial considerations to material ones, one finds that biotechnology is not a labour-intensive technology. Even where biotechnology is growing and developing, it is not employing a large labour force. From the ecological point of view it can be said that biotechnology largely represents a "soft technology", and in particular, applications in recycling and purification may well be labelled as "ecotechnology." Biotechnology can also be "decentralized" into fairly small units, as in the case of bio-gas technology and this points to useful applications at the "village level" in several of developing countries.

As a policy issue biotechnology is flexible and multi-optional; in response to several problems it offers many opportunities.

Potentials

The potentials of biotechnology have been listed in a large number of recent reports and reviews. Every classification of the identified potentials is artificial and is likely to be controversial. Therefore, it would be

perhaps sufficient to discuss the potentials of biotechnology in the context of the present, and the immediate future.

Biotechnology does not lack a history. "Prepasteurian" biotechnology dealt with the production of wine, beer, bread, cheese, the retting of flax, and probably the preparation of pickles. In 1857 Pasteur discovered and explained lactic acid fermentation. This year can be taken as a milestone indicating the advent of modern biotechnology. On the other hand the claim that the milestone should be moved to the 1940s − the era of the submerged production of penicillin − also has its merits. However, in the context of the potentials at our reach today, the historical background is of virtually no interest.

The potentials of biotechnology relate to five F's: fuel, fertilizers, food, feed, fibre, and, with an excuse for orthography, to a sixth one: pharmaceuticals. Fibre is more or less an historical issue, and fertilizer relates to Rhizobium biotechnology. food and feed also have an historical tradition. However, the significant new advancement is SCP (single-cell protein) or the production of microbial biomass for its extremely high protein content. SCP is an obvious bulk product which competes on world markets with protein sources such as soybean produced by traditional agriculture. It seems that SCP has not competed very successfully (possibly soybean protein prices have been regulated to meet the SCP competition), and the interest for SCP has diminished considerably in recent years. The USSR, however, constitutes an exception, since it has announced a programme to expand its production of SCP eight- to tenfold in the next few years.

Further advances in the food and feed sectors concern enzymes and colours. Microbial rennin is already in common use in dairy technology. Other enzymes have also been introduced to food technology: enzymatic tenderization of meat is just one example worthy of mention. Several chemicals used as colouring additives in food technology have been prohibited by recent legislation in many countries. This has stimulated a search for "natural colours" to replace those banned. Microorganisms are efficient producers of, for example, carotenes, and, most probably will be a useful source of colouring agents for the food industry in the future.

In the sector of fuel, biogas has been an issue that has been intensively discussed and reported on in recent years. Biogas production achieves two significant goals: waste recycling results in an acceptable and useful product; and fuel-economy is improved. The first applies

especially to the industrialized countries that face major waste problems (e.g. in Finland there are already farms which produce a major part of their electricity from agricultural wastes through this process). The second goal applies to developing countries where lack of fuel in many areas is a more serious constraint to development than the lack of food.

From the economical point of view, biogas technology represents progress. Untreated agricultural wastes are decomposed anyhow, in and by nature, releasing enormous amounts of energy that are lost as heat. In the biogas production process the waste decomposition is controlled; the methane produced under anaerobic conditions is collected, and utilized to produce heat (for preparing food or for other productive purposes), or perhaps even to produce electricity. The main feature of such technology of interest to the developing countries is the in-situ approach. Production of biogas can take place where the wastes are, thereby obviating high transportation costs to large plants elsewhere. Biogas technology therefore represents a decentralized "village technology", flexible enough to meet varying needs.

Now for discussion of the orthographical insertion — pharmaceuticals. The early post-war years, the late 1940s and early 1950s witnessed a breakthrough in biotechnology — the rise of antibiotics. Within the space of a few years, antibiotics had challenged ethanol (common alcohol) as the major biotechnological product. The era of antibiotics was revolutionary, not only because the substances were now produced in amounts considered unobtainable earlier, but also because of the improvements of the production techniques. Submerged propagation instead of clumsy and inefficient surface cultures became a symbol of this new area. On the other hand, antibiotics constitute a good example of the "rise and levelling off" model. Both products and processes already existed in the discoveries of penicillin, streptomycin and the tetracyclines. What happened was merely an improvement of the process, and certainly improved, it allowed the large-scale production of principally similar product. Today interest in antibiotics has waned. Present-day production of antibiotics is too costly. A cheaper alternative would be the biotechnological production of vaccines and/or antibodies, which processes are possible through the application of genetic engineering. If so, a once outstanding scientific revolution is about to be outdated within half a century.

What is exciting today is the potential of medicine opened by genetic engineering. Substances like insulin (which have been used in therapy

for decades) and interferon (not yet used in common medical practice) will in the near future be available (thanks to biotechnology) in unlimited amounts. Insulin and interferon are just starting grounds. Similar techniques allow for the production of any drug with potential medical use. This is the most remarkable biotechnological potential today.

There exists also another remarkable biotechnological potential — whole enzyme technology. Discoveries towards the end of the last century concerning the separation and release of biological activity from intact living cells indicated that such activity was due and bound to enzymes. Present-day "high technology" permits various treatment of enzymes without interfering with their activity, but improving their usefulness. Enzymes are the real "biological agents", the units which do the "dirty work". Genetic engineering can only help in the mass production of enzymes either as elements of living cells or in a "free" state, but it is the enzymes that represent the ultimate biological potential of such research.

The selection of "domesticated" micro-organisms is still very limited. Though perhaps a joke, it is often said that more than 90% of micro-biologists now active are working on *Escherichia coli*. Translated into biotechnology this estimate reveals that the group of micro-organisms whose potentials are explored thoroughly is indeed very limited. There-fore, much of the future involvement in biotechnology will concern other groups of micro-organisms such as the photosynthetic bacteria, the anaerobes, and the autotrophs ([1]).

The process versus product issue can also be raised in this context. Does not the introduction of new groups of micro-organisms to biotech-nology indicate merely an unfruitful process-improvement approach? The answer is no. Photosynthetic technology would be a substantially new technology in which microbiological methods combined with photosynthetic behaviour could give rise to new biotechnological products.

This is also true for the anaerobes. Life without oxygen is strange to the human being. Life which requires the *absence* of oxygen is even more strange, yet some bacteria exhibit this kind of life. From a biotechnological point of view anaerobic conditions do not constitute a particular problem. Whilst aerobic systems have usually to be effectively aerated, anaerobic systems are more or less self-regulatory. Once established, anaerobic conditions tend to prevail. Much more exciting is

what these anaerobic organisms are able to do. The anaerobes constitute a completely new adventure. In addition to the already cited example of biogas there are certainly other substances to be produced. The "basic fuel", hydrogen, is one of them.

The autotrophs are strange in that they gain the energy they need from the oxidation of various inorganic compounds: NO_2 can be oxidized to NO_3 by some organisms, others oxidize Fe^{2+} to Fe^{3+}. Biotechnologically these abilities are now utilized only in the so-called leaching of low-grade ores. The interesting metals may occur in the ores in a reduced insoluble state. Oxidized iron (Fe^{3+}) may cause an oxidation of the metal in question (e.g. copper) which brings it into a soluble and thus enrichable state. This implies, however, that Fe^{3+} is reduced to Fe^{2+}, and thus loses its oxidative power. As soon as all Fe^{3+} is consumed the process stops. At this crucial point the autotrophic bacteria express their impact. Certain organisms, called thiobacilli, can gain their energy from the Fe^{2+} to Fe^{3+} oxidation. This means that these bacteria, if present, continuously regenerate Fe^{3+} which thus keeps the leaching of the "interesting" metal going on.

Metal leaching is, however, just one application. Further autotrophic applications to biotechnology are, however, only poorly explored. Possibly there exist in the "authotrophic domain" several further potential applications.

The examples given above may show quite convincingly that future applications of biotechnology will utilize new groups of organisms. The search for new products becomes approachable through the metabolic machinery of these organisms. Advancement of genetic engineering will perhaps contribute to a transfer of interesting properties from strange organisms to those that are more easily manageable. However, anaerobic habits and metabolism cannot be transferred to aerobes, nor does it matter, for that transfer would change aerobes to anaerobes. The same concerns autotrophs and phototrophs. Autotrophic and phototrophic behaviour is what researchers have to become acquainted with and to learn to utilize, whether or not this behaviour is expressed in the original autotrophs or phototrophs or has been transferred to some other organisms. Consequently, we are again back at the base — biological information and its expression, and the means of its utilization.

Biological information is arranged in a binary manner: oxidized/reduced; saturated/unsaturated, AT/GC, etc. The biological machinery

thus operates in much the same manner as a computer. This fact has stressed what is now called biocomputers in which inorganic materials presently used could possibly be replaced by organic materials. In fact these biocomputers would be no more "living" than the present ones, but by copying biological logics could be made perhaps more exact and still more rapid than the non-biological systems. This is just one sophisticated example of future possibilities in utilizing biological material. Even with more "crude" goals and approaches, microbiological materials constitutes an almost indefinite source of possibilities.

Prospects

In the news media biotechnology has in recent years been presented as something of a modern "Eldorado." Undoubtedly it has looked like that in the minds of capital investors, because there has occured a veritable boom in the field, and new firms and companies have been established to an almost limitless degree. A "realising factor" behind this boom is, of course, the rapid progress in the field of genetic engineering (or gene technology). In this field in particular it has been difficult, at least for laymen, to separate illusion from reality, or possibility in the distant future from what is possible here and now.

In order to make such a separation possible, biotechnology in the text that follows is described as a sum of its elements, with a discussion of a recent forecast, and in context of the problem of illusion and reality.

According to the definition discussed earlier, biotechnology is the application of scientific and engineering principles to the processing of materials by biological agents (living cells and enzymes) to provide goods and services. A simple example may suffice to indicate what this exactly means. Cheese-making is a biotechnological process wherein milk is changed to cheese. Two key elements of biotechnology are obvious: the feedstock or raw material which is milk, and the product which is cheese. In between these elements is the very process: the rennet which causes the curdling of the milk, the succession of bacteria which (in Swiss cheese) first ferment the lactose to lactic acid and then convert lactic acid to propionic acid and carbon dioxide, the proteolysis which contributes to the cheese aroma, and the proper conditions that are needed for both rennet and bacteria to "work", i.e. the "machinery" which catalyzes the process.

In the case of cheese the product is definite. It cannot be replaced by

a better product because cheese is just what the consumers are looking for. The raw material is also fixed: most cheeses are produced from cow's milk and to a lesser degree from the milk of sheep, goat or buffalo. The properties of milk, the raw material, can be slightly modified; protein or fat content can be increased or decreased by altering feeding rations or by breeding, but milk remains milk, and cheese remains cheese only if produced from milk. The only element or factor that can be significantly influenced is the "process box" between the raw material and the product, in between milk and cheese. Much has in fact already happened in this "box".

Thirty or fifty years ago work in the dairies was done manually (e.g. the continuous mixing of the cheese curd) and the rennet was of animal origin. Now animal rennet tends to be replaced by enzymes of microbial origin and the rather hard manual work is replaced by machines which are controlled and regulated by microprocessors. The bacteria to be added to the milk are carefully selected to be most efficient, phage-resistant, temperature-tolerant, etc.

Swiss cheese (as an example) is expensive because its ripening process is slow, and because it needs long storage under controlled conditions. Further process improvement could involve speeding-up of the ripening process, which in turn would need "faster" bacteria. Faster bacteria could be "created" by some manipulation of the genomes of the involved lactobacilli and propionic acid bacteria, or perhaps more successfully by transferring the useful properties of these organisms to "easier" and by nature faster bacteria.

This example indicates, to some extent, the complexity of even a rather simple biotechnology process. The elements are (1) the raw material, (2) the biological agent, (3) the machinery or framwork where the biological agent acts on the raw material, and (4) the product. The goals are twofold: to maximize output and harvest of the product (a significant element of biotechnology is the separation and purification of the product). In the cheese-making process this takes place when the cheese curd is separated from the whey; in other processes such as the production of antibiotics the separation-purification procedure is much more delicate; chemical means are necessary to minimize the production costs per unit of product.

The prospects, illusionary or real, can and do concern all of the above cited elements. However, illusionary or real projections usually concern only the products (and to some extent the biological agent

following the advent of the era of genetic engineering). This conclusion does not contradict the "technological aim" defined earlier since a projected prospect normally concerns only already well known products (and not the new ones); and how to produce them in higher amounts and at a cheaper, more marketable price. The other elements of the biotechnological process are certainly also worth of consideration. Raw materials are not as fixed as in the cheese-making example. The engineering aspect, that is the machinery or framework and its operations, seems to have suffered from lack of interest with the increased emphasis on genetic engineering. The regulation and control of the whole process is a key element made possible by modern electronic facilities and increased knowledge of systems science. Possibly real economic outputs from developing regulation and control need only modest inputs as compared with developments in genetic engineering.

And the biological agent itself? Almost all current discussion concerns the prospects of genetic engineering. One illusion has to be crushed in this context. Genetic engineering does not create completely new types of organisms. Engineered micro-organisms are *still* micro-organisms. What, then, is still important, and perhaps more important than ever before, is to increase the knowledge about how micro-organisms behave. Even the most sophisticated methods of genetic engineering can never replace the need for fundamental knowledge of microbial physiology. This is a key issue which has unfortunately, been overlooked in most recent reviews of biotechnology, which is wrongly considered to be synonymous with genetic engineering. One significant exception is worthy of mention. The Japanese language does not recognize the concept of biotechnology. Our Japanese colleagues are still interested in the biochemistry and physiology of micro-organisms. This reliance on fundamental basic knowledge may make a considerable difference some ten or twenty years ahead.

The Irish National Board for Science and Technology has published an interesting forecast on "biotechnology trends" ([11]). The study, which was performed by an application of the Delphi technique covered chemicals, pharmaceuticals, health care, and food processing. In a five-year perspective this study indicated a 0.5 per cent median probability that immobilized α-glucosidase will become commercially established for controlling the level of glucose in fermentable substances, that enzyme systems will be used routinely in clinical laboratories for direct measure-

ment of biological parameters (such as blood glucose), that microbial interferon will be marketed by pharmaceutical companies and that microbial insulin will be similarly on sale.

In a ten-year perspective, and with a 0.7 per cent median probability, the study concluded that a commerical process will be developed for the conversion of whey lactose to a sweetener syrup by enzymatic breakdown; that ten per cent of alcohol produced in Europe will be obtained from whey lactose; that the output ethanol produced by fermentation in Europe will increase ten-fold; and that improved methods of enzyme extraction and recovery coupled with enzyme induction in micro-organisms will lead to considerable price reduction and greater supply of intracellular enzymes.

A twenty-year perspective does not add new topics, it only increases the median probabilities of those topics already mentioned.

This forecast is probably realistic; it is at least not overoptimistic. It can be seen that, in a five- or even twenty-year perspective, the list of new products to be produced by biotechnological means is not very long. Forecasts *are*, however, forecasts, and possibly some unforeseeable breakthrough may change the picture significantly. However, whether such a *deus ex machina* may occur or not, the forecast shows that biotechnology is hard work, not a lottery.

The above text implies that there are no short-cuts in biotechnology. The hard facts can be repeated in summary form thus.

(1) There occurs a vast selection of raw materials, particularly materials otherwise considered as waste, that are suited for microbial transformation;

(2) There occurs a vast selection of micro-organisms fitted already by nature with manifold abilities, abilities that can be transferred to micro-organisms which are more suited for biotechnological utilization;

(3) Micro-organisms express their abilities (do their work) maximally in an optimal environment. This environment can be created by engineering, and the activity of micro-organisms in their optimal environment can be steered and regulated by the means of electronics and system science;

(4) The above conclusions give the framework within which known "goods and services" can be produced more beneficially than before. Real and far-reaching progress is, however, bound to completely new products, and completely new goods and services. These are not outside the reach and scope of biotechnology but they can hardly be foreseen.

The restriction of any forecast is due to the fact that forecasting can only deal with future development of present realities; future realities always remain unknown. The Renaissance could not foresee Newton's *Principia,* the French Revolution could not foresee the steam engine: the *fin du siècle* generation could not foresee Planck's quantum mechanics or Eistein's theory of relativity, and the Second World War generation could not foresee space technology, microelectronics or genetic engineering. In our evaluation of the dimensions of biotechnology we are strictly bound to present-day realities.

BIOTECHNOLOGY AND SOCIETY

"The renaissance of biotechnology" is the title of a recent article ([2]). It is well founded because biotechnology is almost as old as mankind. What we are now witnessing is a renaissance, a re-evaluation of old knowledge and tradition in the light of consistent new information. What has been more or less a handicraft for centuries is now becoming a technology. This background of tradition means wealth. In fact one of the most significant credits of biotechnology is that it relies on an infinite tradition, and is, therefore, familiar to and acceptable by the general public.

What has been emphasized above implies that in one way biotechnology is already implemented in society. This is the basis to build on since any prediction of the future must be founded on experiences in the past. Looking at the biotechnology/society interface in the past, several examples are self-evident. The first is certainly vaccination. We cannot, of course, clearly realize what smallpox and the fear of smallpox meant to society some 200 years ago, and our imagination of the impact of the new vaccination technique in that society is similarly vague. It is quite obvious, however, that vaccination introduced a change in values and thoughts. Still in the childhood of the one of the present authors vaccination against smallpox was needed to gain full "civilian rights"; if the

certificate was lacking for some reason, you were merely a second-class citizen. This seems even more interesting since we now know that one inoculation does not guarantee life-long protection and that the inoculations of the ancient times were, in fact, merely social demonstrations rather than medically exact.

Social demonstration or not, smallpox has almost completely disappeared from the industrialized countries as well as in the developing world. This is most probably due to a concerted social and biotechnological (production of vaccines) impact.

Another, perhaps trivial, example is provided by antibiotics. Every student in microbiology is aware of Paul Ehrlich's search for the "chemical knife" or the "magic bullet" — the effective chemotherapeutic substance. To some extent the impact of antibiotics is perhaps not as deep rooted as the impact of vaccination — but one may be biased by the much shorter perspective. In fact, it is already demonstrable that diseases, which, only one generation ago, were feared as potentially lethal — scarlet fever is one example — are now considered as relatively trivial.

A further example of the social impact from the side of biotechnology in the non-medical field is sewage purification. With increased urbanization and increased water consumption there occurs in many regions of the world a shortage of potable water. Water has to be recycled for continual re-use. The technology of purification is of primary concern for efficient recycling and in its technology the biotechnological element is of utmost importance. This biotechnological impact is perhaps not expressed as dramatically as in vaccination or in efficient chemotherapy, but one can imagine what would happen if biotechnological water purification were suddenly to stop.

A most important application with obvious social repercussions may be cited from the field of food technology. Francois Appert invented in the early 19th century a method to conserve food by heating in closed vessels. The original method, although honoured at that time, was clumsy and impractical. It defined, however, an idea (it described, in fact, a completely new product) which, six decades later (in the 1870s), led to the invention of the autoclave. With this tool at hand the canning industry rapidly developed and this development influenced not only the customs, values and health of ordinary families, but also the food market, transportation systems, storage space demands and so on.

Further examples could be cited. They would not, however, alter the picture. One significant feature of this picture is the long time-span —

the inventions of vaccination and food canning are still the cornerstones of modern-day epidemiology and food technology respectively took place around two centuries ago. This implies that a comprehensive evaluation of the social impact of, say, genetic engineering may not be possible in the immediate future. In this respect the science and technology politicians of the next century will be better informed than their colleagues at present.

In accordance with what has been stressed in previous sections, that any evaluation *a priori* can only deal with discoveries, inventions and trends already known, evaluations of future social implications remain merely guesses. The evaluation *a posteriori* identifies the really "new" breakthroughs of the past, but is, of course, valueless except as a lesson to learn from.

One further point may need some attention. It has been stressed ([13]) that one characteristic of scientific and technological progress is confrontation with moral and ethical norms and values. If progress is significant and extensive enough our norms and values may adapt themselves to the change implicated by scientific and/or technological progress. This *problematique* was touched upon earlier. The social impact of biotechnology is obvious in this context: i.e. secrecy of R&D (either for national security or commerical reasons); R&D as an immaterial property which can be owned; micro-organisms as a material property and, the limits of life manipulation.

These, and several others, are problems which effect our society and to which our society now has to react. Since one cannot escape these problems it is better to accept them and to try relevant solutions. It is not even a question of solutions only. The biotechnology/society relationship is a two-way process. There is not only an impact of biotechnology on society, but also an impact of society on biotechnology. The future society is perhaps to some extent influenced and formed by biotechnology. Where society is able to formulate its needs and expectations in clear terms, science and technology (also biotechnology) will react. The contraceptive "pill" (and coil) and antibiotics are good examples of fulfilment of a societal need rather than of old-fashioned expectations. Society has by "trial and error" almost solved the problem of breeding domestic animals. Today, modern genetics provides an outstanding tool for still better results.

The two-way relationship, of impacts and expectations, between science and technology on the one hand, and society on the other,

represents a mechanism which cannot perhaps be very efficiently steered and controlled, but which incorporates much human common-sense.

CONCLUSION

An *a posteriori* evaluation of the socio-economic impacts of biotechnology draws attention to the flexibility and the capacity of biotechnology, and to the wide dispersion in society of the "outputs" from biotechnology.

The flexibility is obvious from the application of biotechnology for such very different purposes as, for instance, leaching of metal ores and the production of interferon. The interferon example is also useful to show the extent of the capacity of biotechnology. Non-biotechnological methods have made it possible to produce only very limited amounts of interferon for scientific use, but not enough for extensive clinical trials. Biotechnological production of interferon will represent an extensive capacity which will open new possibilities both for R&D on interferon and its applications in clinical medicine.

The wide dispersion of the results of biotechnology implies that "goods and services" created by biotechnology (vaccination, easier care of diseases, pure water, and safe food, just to recall examples referred to earlier) are within the reach of every citizen.

Flexibility, capacity and broad impact (in the senses used above) are features which, within the framework of administration, concern various authorities (public health, trade and industry, food and agriculture, public works, and last but not least, education and research). This "general involvement" of biotechnology calls for national strategies in the field. The demand of such strategies has been further underlined by the expected breakthrough of genetic engineering. Genetic engineering has brought new types of problems such as laboratory and research security regulations, applicability of existing common sense and legislation as far as immaterial property is concerned (can a method to cut a DNA-sequence be owned and if yes, by whom?), and the ethical limits to exploitation of the resources of nature.

A number of biotechnology reports from various countries are available; reports have also been compiled by international organizations, governmental and non-governmental. Conclusions vary to some extent, depending on local conditions and national economy, but in

general they relate to the same problems (and similar propositions) which are identifiable in the different reports. Development of biotechnology and the increase of national capacities as related to biotechnology is considered without exception as a significant goal. As obstacles to be removed, the reports frequently draw attention to improper educational curricula, limited research facilities, inefficient organization and management of research and in particular to insufficient cooperation and coordination between the public and private sectors. As regards education, it is worthwhile to mention that education in biotechnology is too broad in scope to allow development of all of its sectors (at least not in small economies). Therefore, selection of priorities has to be made, and these selections should be based on economical realities (such as applications in already established branches of industry, or "new products" which fit the resource base as well as obvious national and international markets). The research-intensity of biotechnology is strongly stressed, which puts particular emphasis on immaterial "know-how" as a substantial element and also as an output of biotechnology and as an object of trade. As further objective for an advantageous and successful strategy in the field of biotechnology, efficient transfer of technology, international harmonization of biotechnological regulations, improved access to relevant data bases and better utilization of bio-informatics, should be considered in depth.

It has been stressed repeatedly in this text that biotechnology produces both goods and services. In addition these goods and services occur and are utilized in several sectors of society. Therefore, gross economic impacts are extremely difficult to estimate. The recent OECD report (1) compares production values as shares of GNP for some "biotechnologically related industries" in a few OECD countries. Food industries contribute, according to these figures, to some 15 per cent to the GNP in New Zealand, as compared with only 7 per cent in the Federal Republic of Germany. As to chemical products the shares are around 15 per cent in the Federal Republic of Germany and Japan, but only 6 per cent in New Zealand. Drugs and medicines contribute considerably less, ranging in from 1.4 per cent in Japan to 0.2 per cent in Norway. These figures are, of course, only crudely indicative. Although the production is "biotechnologically related", it is not possible to identify, at least not in statistical figures, the "biotechnological element" in various products. It is certainly higher in wine than in frozen shrimps, but this conclusion gives no basis for comparing the

impact of biotechnology in the national economies of, for instance, Iceland and Spain.

The technology forecasts have been discussed in a perhaps somewhat provocative manner. The claim was that forecasts are based on existing knowledge, and cannot, hence, predict the really new ideas, which constitute the true scientific and technological innovations. On the other hand, values and expectations of society have a long span, and since society exerts an impact on science and technology, types of goods and services which will be searched for can be defined with some certainty.

In the framework of existing knowledge and "looked for" goods and services progress in the form of products produced by biotechnological processes can be expected in the pharmaceutical industry (in addition to human insulin and interferon, hepatitis-B vaccine, and human growth hormone that are most frequently mentioned; and, predictions for introductions of industrial production do not go beyond this decade). Agriculture is another field where progress is possible; however, in a longer perspective, *nif-* and *hup-*gen transfers to plant chloroplasts are not likely to succeed in the near future (at least not at the level of application in practical agriculture). Microbial insecticides representing a useful innovation are already in use, although outdoor field application needs improvement. Active lignolytic non-cellulolytic micro-organisms would be extremely useful in the wood-processing industry. "Photosynthetic technology" (which utilizes photosynthetic bacteria, particularly cyanobacteria) is likely to progress even rapidly in a foreseeable future.

Progress which corresponds to obvious needs in society probably has almost immediate socioeconomic effects (examples: the new contraceptives, and information technology). A successful transfer of the *nif-*gene to plants, and the successful cultivation of those plants in agriculture would probably have revolutionary effects. Those effects are, however, not foreseeable in any details.

The examples and impacts discussed above concern so-called "high technology" whose implementation and successful utilization requires a well developed cultural, economical and technological infrastructure. Transfer of such "high biotechnologies" to less developed countries is, therefore, not likely to be successful. In an analysis several years ago Gyllenberg and Pietinen ([14]) concluded that "softer" and less sophisticated biotechnologies, such as control of food deterioration, food processing by fermentation and the increase of nitrogen-fixation, correspond to basic needs in developing countries on the one hand, and are

practically applicable under existing conditions on the other. This conclusion may still hold, in principle, and is in line with the policy of the Unesco/UNEP/ICRO MIRCEN project. As infrastructural obstacles are removed more complex technologies can be introduced, but significant policy issues for development, are still small units, simple equipment, independence from heavy transport facilities, labour-intensity, and management and control of the process by experience rather than by formal training.

It can be generally concluded that the various policy reports attach much importance to the products, the goods and services to be accumulated by biotechnology, and to the biological agents which will make that accumlation possible (genetic engineering) but less so to the raw materials and to the framework were processes take place (the engineering and downstream aspects). Except for genetics not much interest is shown in other underlying basic research efforts. It seems, therefore, as important to stress the necessity of strengthening research in microbial physiology and ecology in order to assure that basic biological prerequisites are met in attempts to apply biotechnology in practical agriculture or industry. As far as agriculture is concerned, the need of a better understanding of plant physiology has been stressed. Further priorities of the present and the future should be the development of automation, that is computer optimization, regulation and control of bioprocesses. Considerable progress has occured in this field but in spite of the tremendous development of automation in general, applications to biotechnology are still limited. Hence there occur possibilities for successful inventions, and significantly improved results.

The significance of the adoption of the submerged fermentation technique to large-scale production of antibiotics has been emphasized. Another technical development of almost comparable impact was certainly the introduction of continuous culture (or fermentation), which was developed in the late 1950s. So-called "solid-state fermentation" is probably a technical invention of comparable dimensions. The solid-state technique, now under rapid development, will open new possibilities in, for instance, waste treatment (successful applications are already found in the use of straw as raw material for SCP or alcohol production). New possibilities for solid-state biotechnological waste treatment will, most probably, increase the use of various wastes as raw materials for biotechnological processes as well as increase the impact of recycling technology as something more than a bright idea.

It remains to be concluded that biotechnology, as it now develops,

also faces ethical barriers. Some policy reports speak frankly about the "Faustian bargain", a concentration of the interest of academia with those of business enterprises. Reference has also been made to industrial lobbying which has perhaps caused a dissolution of earlier concerns about the ultimate risks of genetic engineering. So far it seems, however, that the risks concern merely ideologies not practices. The risks are moral and ethical, not (bio)technological. Still, they concern biotechnology and biotechnologists.

REFERENCES

[1]   Anderson, A. (1984). 'Search Begins for Superbugs'. *Nature* **310**, 172.
[2]   Bartells, D. (1984). 'Secrecy in Biotechnology is Short-Sighted'. *Search* **15**, 183—184.
[3]   Budiansky, S. (1984). 'Rows Continue over US Patent Life'. *Nature* **310**, 176.
[4]   Bull, A. T., Holt, G. and Lilly, M. D. (1982). *Biotechnology-International Trends and Perspectives.* OECD, Paris, France.
[5]   Ministry of Education (1982). *Biotechnology in a Small Country – Finland as an Example,* 10 p. Helsinki, Finland.
[6]   Foster, R. N. (1982). 'A Call for Vision in Managing Technology'. *Business Week*, 24 May.
[7]   Andrews, F. M., (ed.) (1979). *Scientific Productivity,* Unesco Paris, France/ Cambridge University Press, New York.
[8]   DaSilva, E. J. (1984). 'Microbial Technology: Some Aids and Barriers in Development'. *Journal Intercultural Relations,* **8**, 413–432.
[9]   Nelkin, D. (1982). 'Intellectual Property: The Control of Scientific Information'. *Science* **216**, 704—708.
[10] Adler, R. G. (1984). 'Biotechnology as an Intellectual Property'. *Science* **224**, 357—363.
[11] National Board for Science and Technology (1982). *Biotechnology Trends.* Pp. 21—24, Dublin, Ireland.
[12] DaSilva, E. J. (1981). 'The Renaissance of Biotechnology: Man, Microbe, Biomass and Industry'. *Acta Biotechnologia* **1**, 207—246.
[13] Gyllenberg, H. G. (1985). 'Progress and Expectations in Biology'. In *The Identification of Progress in Learning,* T. Hägerstrand (ed.), Cambridge University Press, Cambridge, U.K.
[14] Gyllenberg, H. G. and Pietinen, P. (1973). 'Future of Microbiology in Developing Countries', Report No. 5, 44, Dept. of Microbiology, University of Helsinki, Helsinki, Finland.

Department of Microbiology, University of Helsinki,
Helsinki 71, Finland

and

Division of Scientific Research, and Higher Education, UNESCO,
Place de Fontenoy, 75007 Paris, France.

INDEX OF SUBJECTS